PELICAN BOOKS

A 629

THE SCIENCE OF ANIMAL BEHAVIOUR

P. L. BROADHURST

THE SCIENCE OF
ANIMAL BEHAVIOUR

P. L. BROADHURST

—

WITH EIGHT
PLATES

PENGUIN BOOKS

Penguin Books Ltd, Harmondsworth, Middlesex
U.S.A.: Penguin Books Inc., 3300 Clipper Mill Road, Baltimore 11, Md
AUSTRALIA: Penguin Books Pty Ltd, 762 Whitehorse Road,
Mitcham, Victoria

—

First published 1963

—

Copyright © P. L. Broadhurst, 1963

—

Made and printed in Great Britain
by Cox and Wyman Ltd
London, Reading, and Fakenham
Set in Monotype Bembo

To my parents

Contents

Acknowledgements

THE following acknowledgements are made for the use of photographs: Routledge & Kegan Paul Ltd for Plate 1 (upper photograph), from *Experiments in Personality* (ed. Eysenck), and for Plate 7, from *The Dynamics of Anxiety and Hysteria: An Experimental Application of Modern Learning Theory to Psychiatry* (Eysenck); Plate 2 (lower photograph) by courtesy of Dr Bernard Weiss and reprinted from *Science* by permission; Plate 3 by courtesy of Dr Harry F. Harlow, of the University of Wisconsin, and the Editor-in-Chief, *American Scientist*; Plate 4, Three Lions Studios, Inc.; Plate 5 by courtesy of Dr Harlow; Plate 6, by courtesy of Thomas D. McAvoy and *Life* Magazine 1955 *Time* Inc.; Plate 8 by courtesy Mr Leonard Kamsler.

Editorial Foreword

PEOPLE sometimes ask: what has the study of animal behaviour to do with the 'proper study of mankind' – human psychology? A part of the answer, at least, will be found in this book.

True, animal behaviour is, as it always has been, a fascinating study for its own sake. It has been such since before the time of Aristotle; and since Darwin it has been found a rewarding pursuit by many great naturalists. In the writings of the naturalists important observations were not infrequently mixed up with faulty, 'anthropomorphic' interpretations. During the present century there has been a great advance in the development of experimental and other observational techniques, and in the formulation of rigorous principles of interpretation. Today there are two main fields of research concerned with animals. There is 'ethology', in the main the study of the behaviour of animals in their natural environments, and there is the study of animal behaviour in the laboratory. In this book the author has given some account of the studies of ethologists but fears, he says, that readers may detect his 'laboratory bias'. If bias there is it is understandable and forgivable. It is more than forgivable when an author prefers to write most about the things he knows best and at first hand. Dr Broadhurst is an authority on laboratory experimentation with animals. (He has been engaged continuously in such studies for over ten years and within that time has published many important scientific papers. This Pelican is No. 30 on his list of major writings and there is another thirty which he modestly excludes from this list.)

But while working in the laboratory he has had time to look out of the window, and he has *made* time to think about the future of man. In some ways his book reads like a detective story as do so many good scientific books (after all, what is scientific research if not a kind of detection?). But it differs from a detective story in one important respect. In reading a 'whodunit?' we must not look first of all at the last chapter: that is cheating. With this book there is much to be said for having a first quick look at the last chapter. Here the author reviews some of the astounding implications of animal psychology for human society – the possibility, for example, that pigeons might pilot planes and that chimpanzees might be engine-drivers. Read by pessimists this chapter might suggest that the future of the skilled worker is indeed dim. When he is not made redundant by automation he will be made redundant by the slave labour of pigeons and monkeys. More optimistic readers will take comfort in the thought that in a still more affluent New Jerusalem men will go fishing while their pigeons earn for them their livings. This is a book not only for naturalists and specialists in 'behaviour theory'. It is a book also for 'Top Executives' and Industrial Psychologists who must prepare themselves for developments which are only just round the corner. Chapter Seven is not just science fiction. These things *could* happen here.

C. A. MACE

I

Why Study Animal Behaviour?

EVER since the dawn of history man has been interested in the other denizens of the world around him. According to Hebrew mythology as it comes to us in the Bible, the actions of the serpent in the Garden of Eden played a crucial role in the subsequent history of mankind. Similar stories occur in other mythologies: Greek and Norse mythology are full of tales of divine intervention in animal guise or through the medium of some animal's special characteristics or qualities. The bestiaries – the natural-history books of the Middle Ages – were enormously popular and are full of descriptions of fabulous creatures. Not only were they concerned with the form and shape, that is the morphology, of the creatures they portrayed, but the way they behaved was also the subject of speculation often of a sort incredible to us, and quite devoid of foundation in observable fact. Bestiaries, for example, record that bears lick their offspring, born a formless mass, into shape; or, again, that certain snakes suck the milk from cows. Nevertheless, the gentle unicorn and the evil dragon were as familiar to educated, medieval man as the camel and the chimpanzee are to us all today.

But there is another side to the pre-scientific observation of animals. The practical consequences of the way animals behave have also been a subject of concern since prehistoric time. The domestication of the dog as a companion and guardian for man began before recorded history, possibly in the time of Neanderthal man, a thousand centuries ago; the keeping of cattle, birds, and bees is also of great antiquity. Now in order to domesticate a species of animal it is necessary to breed it for the qualities which are desired: the faithfulness of the dog, the docility of the cow, and so on. These are behavioural characters, that is, they designate not the way the animal looks – though this, too, is often important also for the animal breeder – but what it does. The many varieties of thoroughbred dogs available in the world today are all probably descended from one or perhaps two wild species – the wolf and the jackal – and, in addition to the vast differences between them which are obvious to the eye, there are also the more subtle differences in temperament and character, as well as some very striking differences in the way they respond to various situations. For example, the Dalmatian, or spotted dog, is a coach dog, and was bred to run under coaches in the days before motor cars; the Springer spaniel, on the other hand, could hardly be less interested in things on wheels, but will readily learn to leap into action to retrieve a falling bird. It would be hard to train either species in the other situation, for the potentialities for behaving in a certain way, bred in through many generations, are just not there. Now this selection process depends for its success on the very careful observation of the behaviour which the breeder wants to bring out; and down the ages man has obviously become very expert in doing so.

Of course, it is practically meaningless to speak of behaviour in general, just as it would be to speak of weather in general, without having in mind what is happening at a particular place and at a particular time. Thus behaviour must always refer to a particular animal carrying out some particular act in particular surroundings in response to some particular set of conditions, and it is on the observation of the details of many such occurrences that we base what we know about the behaviour of the whole class of animals of which the observed ones are members. In this way we are able to say, for example, that bats hunt their prey by means of a form of echo-sounding without having to specify all the details of the type of noise they emit when they are hungry and flying around searching for the moths and other insects they feed on, the way their ears hear the sounds reflected back to them, the conditions in which they cease to be able to do this, and so on. All this is implied in the statement that bats hunt their prey by means of echolocation, and this adequately characterizes their behaviour. But echolocation is not an attribute of bat's behaviour, it has no real existence like the ink on this paper, it is merely a shorthand way of thinking about and describing the way some animals behave under certain circumstances. And so it is with many of the terms we shall use in talking about animal behaviour – instinct, discrimination, and the like ultimately refer only to the actual observations of many people looking at the way animals behave.

What then do we mean by 'behaviour' in this context? Since we are not and cannot be concerned with what is going on in the mind of the animal, this is an easier question to answer than it might otherwise be. Consider human 'behaviour' for a moment. Now this might include thinking about the answer to a problem, dreaming, or imagining a new experience – none of which might be detectable to the outside observer except and unless he were armed with delicate physiological recorders attached to the person concerned, and then not always. So animal behaviour is usually seen as some definite movement often in response to some definite event occurring in the immediate neighbourhood (a 'stimulus', as it is called). Most of the exceptions relate to responses to internal events which are known, usually with a fair degree of certainty, to be going on inside the animal – not in its 'mind', but rather in its body, and especially as related to secretion from its glands – the hormones or chemical messengers in the bloodstream. The behaviour we observe is often compounded of many movements, and thus may usually be said to involve the whole of the animal in some way or another. Thus we are not concerned with minute responses of individual cells or even groups of them bound together as, for example, in a muscle responding to stimuli which cause them to contract. This is the province of the physiologist who studies such automatic reflex responses. Our study is of the animal as a whole, responding as it does in many ways to its often rapidly changing environment. This is behaviour study, and this is what this book is about.

It is principally concerned with the way the study of animal behaviour is carried on. This has nothing to do with the casual way people speculate about what dogs or cats are thinking, or even the unsystematic – though often very useful – way animal breeders try to discern the traits they want to emphasize in future generations. What we shall be dealing with here is the scientific study of behaviour – the psychology, in the true sense, of animals. The ways in which this differs from the ordinary view of animal behaviour are numerous, and will become more apparent as we proceed, but they do not vary in principle from the way in which scientific thought and method in any subject, be it trade, the weather, or sex, differ from our casual thinking about it. Perhaps the most important of all the differences, however, and one which must be mentioned right away, is the need for utter objectivity. We cannot see into the animal's mind, any more than we can see into that of any of our fellow humans, but that doesn't prevent us in both cases from thinking that we can. The animal may seem sad or happy or angry, but we cannot infer that this is the case from the way we ourselves might feel in the same situation. To do so is to indulge in anthropomorphism – seeing man's shape in all things – and this is the cardinal crime for the animal observer. It may be that we are right in thinking the animal is sad or happy or angry, but the only thing we can know with certainty is what it *does* – how it behaves. This we can observe, measure, record, and so on, in the ways this book will describe, and this is where the need for objectivity is paramount. The behaviour alone, with no interpretations, intuitions, or inferences, must be faithfully recorded in ways as impersonal as we can make them, so that we can be sure that if two people do the same experiment, or record the same observation, the results will be identical. Speculation and interpretation have of course their place in science, and in the scientific study of animal behaviour as in any other branch of science, but they come later, when the data are all recorded with complete objectivity and conclusions drawn from them are established.

At this point, too, we must pause to consider what we mean by 'animal'. After all, this book was written by an animal – and, incidentally, is being read by one! So strictly speaking we should say 'non-human animals'. But this is too clumsy a circumlocution to use often, and henceforward the word animal will imply the exclusion of the human species. This common-sense usage also tends to exclude birds, insects, and fishes, and this will generally be the case here also, partly because these classes of animals have been less extensively studied than the four-legged warm-blooded mammals, like our domesticated species. These include laboratory animals such as rats and mice which have been deliberately domesticated and, as well as our pets (dogs, cats) and farm animals (sheep, cattle, goats, etc.), some two-legged species such as some of the apes – a group which, it is as well to remember, includes man himself.

But why study animal behaviour at all? A practical application has already been mentioned – the improvement of breeds of domestic animals the better to serve man in his work, and this has often been the reason for much practical study. For example, the behaviour of the hen when it goes broody is of great interest to the poultry farmer – because it stops laying eggs. But there is essentially only one basic scientific interest in the study of animal behaviour and that is to learn more about man himself. The pursuit of knowledge for its own sake is often spoken of but rarely practised in pure form. Thus in studying an animal we are concerned with the way it learns, with its instinctive behaviour, or with the way learning and instinct have evolved, and we study in order to understand human learning, human instinctive behaviour, and how human behaviour has evolved from that shown by man's nearest relatives.

But then, it can be legitimately asked, why do you not study humans themselves? The answer is that often it is not possible to do so, for a variety of reasons. To take the study of learning first. In general it is necessary to provide some incentive or motivation for learning – this can be positive when it is a reward or negative when it is a punishment. Both are used in teaching children – the high marks and the teacher's praise are rewards, low marks, detention, and even chastisement are obvious punishments. And so it is with animals, but with animals it is possible to use greater rewards and punishments than is possible with children, but which are indispensable for the success of the study. Thus it may not be possible to induce the animal to behave in the way desired without giving some very large reward, equivalent, say, to most of its day's ration of food at one go. Now it would be undesirable to say, the least, to use such a procedure with humans. Or again it might be necessary to produce real fear in an animal in order to learn something of its reactions. This might be done by giving it a brief electric shock to the feet. This too is difficult to do with human subjects because of the risk of harming them.

The second sphere in which experimentation with animals is of paramount importance is that of instinctive behaviour. Here the reasons for avoiding the human species are rather different. It is that there is little instinctive behaviour left to be studied in man. We may talk in everyday speech of our 'baser instincts', usually referring to our sexual urges and the manner in which we gratify them or not, as the case may be, but in truth there are two grave difficulties in the way of studying them scientifically. Firstly, there is the Victorian reticence about discussing such matters, which though on the wane in Western cultures is still sufficiently strong to make it difficult for a serious investigator in the field, like the late Alfred Kinsey, to obtain information about human sex life, and wellnigh impossible for anyone to obtain scientifically reliable data. There are reports of electrocardiographic recordings of heart rate taken from human couples having sexual intercourse, but there are none of scientific studies of behaviour. And secondly, the instinctive components of sexual and indeed of any other human activity are difficult to discern. So much human behaviour is learned, as opposed to inherited, that it is almost essential to turn to lower animals to be able to study clear examples of instinct.

A third sphere of animal study – the evolution of behaviour – provides an even clearer example of the difficulty of trying to collect data from humans. The study of evolution of a characteristic – be it physical, like the shape of teeth, or psychological, like behaviour – is inevitably a genetic study. This means that the inheritance must be studied, and the standard way of doing that is to measure the differences in the characteristic in question between different strains of the same species of animal and then to cross them and study the offspring. Another standard method is to select for mating certain animals from a larger population, much as the animal breeders do in trying to improve strains of domestic animals, and then to repeat this for successive generations. Now obviously either of these procedures is unthinkable in human society as it is organized at present. Freedom of choice of mate is so fundamental that any suggestion that it could in any way be curtailed, even for racial improvement, let alone for experimental purposes, would be vigorously resisted. Perhaps there will come a time when human hereditary endowment will be assessable in a form which will permit the beneficial potential in human offspring to be made the best use of by judicious matching of mates, but that is obviously so Utopian as to be beyond the bounds of possibility at present. Even if it were not, the slow reproductive rate of humans would make many experiments of this kind a marathon which the originator would always lose. He could never, in the nature of things, see it to completion. And finally, as if these difficulties were not enough, there is the necessity to control rigorously the environment in which the subjects of any such experiment must be reared. This is also true of the other types of investigation of learning and instinctual behaviour mentioned above, but it is especially necessary in any investigation of genetic effects. This means that the way in which the animals are fed, cleaned, handled, and even kept warm must be identical. Now this is sufficiently difficult to achieve for small animals such as mice and rats as to make this sort of investigation rather rare; to do it with humans, who need so much longer and more varied a period of development – say twenty-five years as opposed to fifteen weeks for the rat – would be virtually impossible. All in all, therefore, the near impossibility of genetic-breeding studies with humans serves as an especially good example of the need to make use of nonhumans.

Hence the recourse to animals. With them it is possible to provide rewards for correct learning or punishments for mistakes of a size, relatively speaking, which would be out of the question with humans, adults or children. In animals it is possible to observe instinctive behaviour which in humans has long since been covered by the veneer of civilized custom, and it is possible to interbreed them in any way desired and to rear the offspring under any required conditions. And yet this privilege, this freedom of action to use any species of animals in any way best suited to the needs of his inquiry, essential though it is to the scientific student of animal behaviour, carries with it both responsibilities and disadvantages.

The disadvantages are more obvious than the responsibilities. 'The proper study of mankind is man', as the poet has it, albeit in another context, and so it is. If we study, say, fear in an animal other than man, how do we know that the conclusions we arrive at are valid for man? The answer, in brief, is that we don't. And yet there are so many congruences between fear in man and fear in animals that it becomes possible to make useful inferences. The differences between man and animal in respect of fear are also known and can be taken into account, as well as providing useful pointers to similarities. What has to be guarded against are differences which are as yet not known; these constitute a real danger which can invalidate conclusions drawn across species. But this is a difficulty of the comparative method in any science. The skeletal structure varies between man and beast, and yet this doesn't prevent studies of, for example, tooth decay in the ferret from having a bearing on the prevention of caries in man. Cancer tumours will grow in mice with the equally unfortunate facility that they display in human tissue, and this provides an important method for studying this disease. And so it is with behaviour. Yet it is often felt that the cases are somehow different: that to jump the gap from animal to man is perhaps permissible with physical characteristics but that, in the case of behaviour and the mental awareness which is thought necessary to accompany it, the leap is too much. Man is unique in his consciousness, it is argued; his power of thought is what distinguishes him from the lower animals. How then can you hope to learn anything about such unique qualities by studying them in animals, which, by definition, do not possess them? There are two points to consider here. Firstly, it is generally agreed nowadays among psychologists and other behavioural scientists that awareness, or consciousness, and indeed all mental processes, are dependent upon physical structures and chemical changes in those structures – that is, the nerves and the brain. Now lower animals have nerves and brains, and the control which they exercise over behaviour is not detectably different from that found in humans. The messages pass to and fro along the different nerves in the same way, the nerves and brain are constructed in much the same way, and the messages which pass through them are of the same electro-chemical nature. So it turns out that it is phenomena intimately dependent upon physical structures, though not the working of those structures themselves – that is the work of the physiologist – which we are studying after all, and in this way we are no different from any other scientists using comparative material from animals. It may still be objected that you can never study the thoughts of the animal in the way you can study the thoughts of humans. This is an outmoded view of what a psychologist does. Psychologists no longer attempt to investigate other people's mental life by asking them about their thoughts. The mere fact of asking the questions is enough to distort the process, and the fact of having to formulate the answer is even more disturbing. It has long been conceded that the only thought processes really open to study are one's own, and the difficulties of being both subject and observer at one and the same time are great and give rise to problems which have never been satisfactorily overcome. Therefore it is not to be expected that methods which are recognized as unsatisfactory and unscientific when applied to humans should be applied to animals.

This is not to say that the psychological study of humans is not largely dependent on speech: it is. But it is speech viewed in a special sort of way. Thus we may ask a question about the way somebody feels about a particular topic or after a particular experience not with the hope of learning about his thought processes at the time, but rather to obtain clues about his past, and often his future, behaviour. The fact that he answers the question in a particular way may be what is important, not the content of the answer itself as expressed in the words used. For example, we may ask a person in a personality test the question 'Do you prefer gardening to playing cricket?' not because we are interested in discovering how he spends his leisure time, but because research has shown that a particular answer, one way or the other, will be an indicator of the behaviour of the person in many other situations. The point becomes obvious when we consider what we are doing when we ask a child 'What does two plus two make?' We have no wish to learn the answer for ourselves – the child's spoken answer 'Four' merely tells us that he knows the relationship. Now, clearly, we cannot do such things with animals, and there is no denying that in many inquiries this is a drawback in using animals instead of humans. But it is usually possible to get round the difficulty by devising some way in which the animal can respond, can show by its behaviour that it knows the answer to the problem you have set it, and in this way the difficulty can be circumvented. For example: you might want to discover if an animal has colour vision, in other words to ask it the question 'Do you detect any difference between these two lights?', one being, say, green and the other, of equal brightness, red. Training the animal in ways we shall be discussing later will enable it to answer, by, for example, always going to the red one to feed, irrespective of where you put it. In this way we can know with as much certainty that it sees the lights as different as we would if we were able to ask the simple question which we would obviously put if we were doing the experiment with human subjects.

The second point to consider in the jump from animals to man is that, though human psychological make-up may be very much more complex than that of any animal, in terms of its potentialities and the amount of information stored, its uniqueness may not be due to some special quality but rather to this complexity itself. That is to say, the elements of the behaviour we display in our very complex civilized lives are – apart from speech – all present in lower animals but in simpler form. Now it is a cardinal principle in science to proceed from the simple to the complex, and so this lack of complexity in animal behaviour, often however more apparent than real, may be of great use. Indeed, if we cannot understand thoroughly the simple behaviour of a lowly animal, how can we hope to fathom the complexities of man?

These then are the major disadvantages of the psychological study of animals. The absence of speech and the lesser complexity can be overcome and indeed turned into advantages in certain cases. What then are the experimenter's responsibilities? These are threefold: to the animals he uses, to himself and his colleagues, and to society in general. To his animals in a laboratory, the experimenter owes a special responsibility: these are sentient organisms which are wholly dependent upon him for their food, water, care, and comfort. They have no voice and no vote; they have no say in whether or not they shall participate in an experiment which may involve some degree of discomfort. But they are not friendless, and both within and outside the laboratory there are groups of people, sometimes vocal and well organized, who are ready to defend them against any abuse which comes to light. But in the day-to-day affairs of the laboratory the experimenter rules, and his duty is clear. It is to respect life, in no matter how lowly a form it appears, to avoid imposing any unnecessary suffering, and to promote the welfare of his animals at all times. Indeed, it is a short-sighted policy on his part to do anything else, for results obtained from experiments on animals which are sickly or scared may at best be contaminated by these factors and at worst be worthless. To himself and his colleagues, the experimenter owes this same degree of care for his animals, though not in this connexion for their sake but for his own and that of his fellow-workers. Animals maintained in insanitary conditions are more liable to disease, and though, contrary to common belief, there are few diseases among the more frequently used laboratory animals which are communicable to man, some do exist and workers have the right to be protected from them. In addition, animals in captivity, especially the larger ones, can be physically danger-ous to those humans in proximity to them, and this risk may be increased by discontent. And finally to society in general the animal worker owes an example of humane consideration for dependent creatures, and there are some institutionalized sanc-tions to enforce it. The animal protection societies mentioned earlier are rightly concerned with animal welfare and are active in seeking to secure it. The excesses of the anti-vivisectionists, who would seek to prevent all animal experimentation and the use of all medical benefits derived from animal research, cannot be condoned, but to the extent that they deplore un-necessarily painful experimentation, unsatisfactory husbandry, and the nefarious practices of disreputable animal dealers who steal pets for supplying to laboratories, they are to be supported. Society has a more powerful sanction in addition: this is the law. In many civilized countries animal experimentation is gov-erned by legislation which stipulates the conditions under which it may be performed. Britain is fortunate in this respect: an Act was passed in 1876 which prohibits all experimentation on vertebrate animals which may cause pain, except under licence from the Home Office. Not only must the individual scientist be licensed and the general nature of the experimenta-tion have received prior approval, but the premises of the laboratory must also be licensed for the purpose and are liable to inspection at any time. There are various degrees of licence, depending on the type of animal to be used, and on whether or not it is to be allowed to recover after anaesthesia. Special regulations govern the use on animals of certain drugs of the curare type, which paralyse the muscles but do not occasion loss of consciousness, and hence give rise to the danger of inflicting pain without any overt signs that this is occurring. Each year the licensed scientists must report to the Home Office the number and type of experiments that have been done. By and large this system has worked well enough over the years and is supported by the scientists concerned. It affords a degree of protection for themselves and their laboratories against the more militant anti-vivisectionists, whose deplorable activities in some countries include laboratory smashing. This protection includes the statutory obligation upon any person wishing to prosecute a licensee under the Act for cruelty to an animal during any experiment to obtain first the permission of the Home Secretary. During the currency of the Act the granting of such permission has never been recorded. This circumstance reflects well both upon the care with which applications for licences are considered and the Act administered as well as upon the integrity of scientists in general who work under its provisions.

2

The Study of Animal Behaviour in the Field

ALL psychology, and indeed all science, reduces to observation. Often this observation is as far removed from real life as the test-tube under the laboratory microscope. In the case of the work discussed in this chapter, it is made in the natural setting of real life. Making observations in the natural setting is of importance in many sciences. In astronomy, for example, it is difficult to do it anywhere else; in some geological sciences there may be few alternatives and in biological sciences in general and in the study of animal behaviour in particular it is often of crucial importance to the understanding of a process to see it at work in nature. The mere fact of taking it into the laboratory may alter its character so radically that it is no longer the same process as before. Often this can be recognized and appropriate allowance made, or the domesticated laboratory animal may be studied in its own right, but it is often the case that the study cannot be made at all in any other than the natural setting. Examples might include the migration of birds over hundreds of miles of land and ocean, the social behaviour of troops of monkeys ranging widely in forests, the courtship behaviour of the immense sperm whales, and the foraging behaviour of soldier ants. Sometimes the natural setting may be modified in various ways to systematize the influences acting upon the animal and to give the experimental observer greater control over such influences and over the site of action of the animal so that it can be the more easily observed. And sometimes it can be modified by introducing more objects into the natural setting for special purposes.

Man has been studying the behaviour of wild animals at large since prehistoric times; not only is it the obvious method with which to begin, but sheer necessity makes the human hunter study the ways of his prey. The drawings of stone-age men on the walls and roofs of caves in France and Spain show in their delineations of their animal prey a perceptive knowledge of the elements of the animals' behaviour. Primitive man, faced with the need for tracking and destroying wild animals, became well acquainted with the habits of the animal and with its probable reactions when roused by attack. These are clearly different problems from those associated with the domestication process in which selective breeding gave control over the hereditary qualities which were important for economic or religious purposes. In domestication the environment of the species is radically altered by conditions of captivity or semi-captivity; at least control over the free-ranging of herds is exercised. But, in hunting, the animal's environment has to be accepted and contended with as it occurs. This also holds true for the modern student of behaviour, who wishes to observe the relation of behaviour to environment in a naturalistic way.

However, this is not to say that a large body of information – much of it false – did not accumulate about the behaviour of bird and beast as a result of casual observation, a product of man's natural curiosity about his own environment.

To take but one erroneous example. The legend of the fabulous phoenix – which consumed itself, nest and all, in a fire, only to rise again from the ashes – is believed to have resulted from the habit of certain species of birds of taking dust-baths to relieve themselves of parasites. Sometimes this was done in the ashes of a fire, and sometimes the ashes were still alive and liable to burst into flame when disturbed. So dramatic a sight as the bird rising up from the flames must have been a stimulus powerful enough to perpetuate a legend. Rustic observation of the habits of birds – to take one correct example – is well known. Cuckoos arrive in southern England after their northward migration around the second week in April, and the rhymes which associate their arrival with the dates of local fairs prove that the medieval bird-watcher knew this as well as any ornithologist today. This naturalistic type of animal study nevertheless has its drawbacks: the very casualness of it sometimes has the effect of giving rise to false interpretations of the behaviour observed. It is rare for so wrong – and so dramatic – a misconception as that of the phoenix to arise, and in the modern setting it is unlikely that it would, but there are subtler errors than that. The danger of anthropomorphism in the observation of animal behaviour was stressed in the previous chapter. It is especially likely to arise in the casual but untrained person observing animals behaving in the field or domesticated animals behaving in their quasi-natural setting in the company of man. Lack of knowledge of the antecedents of an action observed is often likely to lead to a false observation or interpretation of the nature of that action, and usually this interpretation is anthropomorphic – that is, it imputes to the animal human qualities of thought or feeling which there is no way of telling it possesses. Such observations are often referred to as anecdotes, as opposed to scientific observations, and it was to counter them that a rule was laid down. This is known as 'Lloyd-Morgan's Canon' after its originator, and it states that we must never interpret a piece of animal behaviour as the outcome of a higher capacity or power of mind if it can be interpreted satisfactorily as the outcome of a lower one. Now this rule idealistically presupposes a state of affairs which is not yet apparent: we are not in a position to decide exactly always what is 'higher' or what 'lower'. Indeed it is not entirely clear what is meant in this context – higher on what scale? Ordinarily, however, 'higher' is interpreted as referring to degree of evolutionary development, or to individual development of the animal in question, and it is usually not too difficult to apply the rule in a way that makes sense. A classic case which occurred in Germany in the last century is that of 'Clever Hans', a horse which was able to calculate – or so it seemed! Given an arithmetical question it would stamp a foot the correct number of times to furnish the answer. Now investigation by a psychologist, Pfungst, showed that the answer was only correct in the presence of the trainer, von Osten, and so it was deduced that the sequence of events was this: despite the lack of any intention to defraud, von Osten was making unconscious small movements when the right number of stamps of the horse's hoof had been reached, and it was this clue to which the animal was reacting. Now this in itself is quite a clever piece of behaviour for the horse to have learned, but it pales in comparison with what had been claimed for it – that is, the ability to count, subtract, and multiply. In this sense then we can appropriately apply Lloyd-Morgan's Canon and say that this performance is a lower one than the other. Armed then with this rule the naturalist can avoid anthropomorphic mistakes; the scientific observer in the field also is guided by it from time to time in interpreting what he observes.

The methods by which the scientific observer puts himself in a position to see the behaviour he is interested in are various and depend on the species he is investigating and the aspect of behaviour he wants to study. Thus, it may be possible to observe relatively statically, as in the case of watching the behaviour of birds coming to and from their nest, or it may involve a chase through the forest comparable to hunting, as in the case of the observer interested in the movements of large animals feeding in various places. Indeed the process may be closely comparable to hunting with intent to kill, except that the camera replaces the rifle.

But the traditional method of observation involves the concealment of the observer from the animal, and this is usually accomplished by the use of a static 'hide'. In bird-watching, for example, this may be a simple canvas construction on the ground near a nest. It may look rather like the wind-breaks which are to be seen on beaches – open on one side (away from the nest) and with no roof. The hide is left permanently on the site selected, and very soon the birds become adapted to it and ignore its presence as they would any non-moving harmless object. The observer is then able to enter and leave the hide without disturbing the birds. Observation is carried out by looking through appropriately cut holes, either with the naked eye or with binoculars or a telescope.

A hide may also be located off the ground, usually in trees. A tall tree is especially useful in observing large animals which roam freely and which might trample down a hide on the ground. A site is often chosen overlooking some natural feature attractive to the species being studied, as in the technique of big-game hunting known as 'sitting-up'. The hide might be a water-hole, or a man-made feature such as a plantation or a specially devised lure. The technique known as 'provisionization' is used by Japanese workers studying monkey behaviour – food is provided for bands of apes in the same spot repeatedly so that they always come to seek it.

With smaller creatures, such as insects, there may be no need to hide in order to observe behaviour. The human observer is too large to be seen. Recently with the development of skin-diving techniques, the marine biologist has been able to enter and stay in the underwater world of fishes and other sea creatures. A whole new area of observation in the field – or rather in the water – has thus been opened up and is now beginning to pay dividends in discoveries about the behaviour of fish especially. It has been shown, for example, that some individual bass and mullet will, day after day, occupy the same place above rocks covered with water at high tide. A locale which the animal thus occupies and defends against intrusion by others of its own species is called a territory. Such territorial behaviour as shown by fish is understandably hard to establish.

The observation of animals in the field is frequently complicated by their mobility, which makes it difficult to follow individuals with any degree of certainty. Individuals may get lost in the group, whereas they can be identified by small differences of marking, etc., in static study of the sort previously considered. Small groups can sometimes be followed by field glasses over a limited range. The use of helicopters for studying the movement of large herds of game has also been tried. Flocks of migrating birds can be followed by radar. But in general the solution to the problems posed by large numbers of animals moving widely has been to use sampling techniques. The principle involved here is to choose individuals in some way and then to study them more closely than the rest of the population from which they come. We then assume they are representative of the rest and are so able to tell us all we want to know, or, to put it more technically, enable us to generalize to the rest of the population. This method is not of course the special province of behavioural study: the geologist taking a sample of rock from perhaps 20,000 feet down in a boring assumes that the sample is representative of the rock in general at that level. So, too, the physical characteristics of a species of animal – plumage, pelt, or skeleton – is definitively established by the samples of that animal killed in the field and preserved in museums, and the nature of extinct species can be established in the same way from samples, often incomplete, such as the skeletons of prehistoric man dug up from time to time. Now behaviour is of its very nature evanescent; it leaves no fossil record, and indeed it is difficult to record satisfactorily in any permanent way at the time it is happening. It may be objected here that motion-picture techniques can record behaviour completely and permanently. This is true up to a point. The use of such techniques for analysing the details of the sequence of movements in, for example, the courtship display of many species of birds, where the natural movements take place very quickly and are difficult to follow, is invaluable. By slowing up the whole sequence the various elements which go to make up the pattern can be identified and the time taken for each can be measured if necessary. Also the actions of more than one animal can be studied at the same time, by concentrating attention on single animals in the group in turn. Such a record may solve controversial problems, such as whether or not an animal does perform a certain action such as eating the afterbirth.

But a complete interpretation of a piece of behaviour based on the movie camera can be as misleading as any traveller's tall tale. Absence of knowledge of the antecedents – the events immediately preceding those captured on film – may distort the significance of what follows. The record for one individual may be incomplete and may have to be supplemented by that properly belonging to another individual, who may for a variety of reasons differ slightly from the first, and so on. It takes a highly skilled observer to photograph behaviour in the field in a meaningful way, and it can be argued that he is better employed there in visually observing and recording by means of notes, rather than concerning himself with the technicalities of photography. Similar considerations apply to recordings of animal cries and bird song, but there is less strength in these objections, for our auditory memories are even more fallible than our visual ones. Also bird song is a special case, for a rather rigid pattern may be reproduced over and over again so that a well-made tape recording can be of great use.

The technique of following a few individuals usually involves marking or tagging them in some way. Then, even if these individuals are hidden by other animals in the group or if they move far and quickly and so are lost sight of, there is a hope that they can be identified again later when they are caught.

The most obvious example of this is the trapping and banding of birds. Analysis of the pattern of the locations from which the ringed specimens are recovered tells us a vast amount about journeyings the birds have made in the interim and so shows the paths migration has taken. Similarly, the movements of fish can be documented: a plastic or metal tag is attached to the fish, which is then released in the same way as ringed birds are; fishermen or others catching the fish and so recovering the tag are asked to report the location to the experimenters, who are again enabled to plot movements. The technique can also be used with insects: individual bees in a hive can be marked with a fluorescent paint so that their comings and goings can be identified among the mass of their fellows, and houseflies can be marked with radioactive elements which betray their presence to a Geiger counter. Small mammals such as voles which tunnel underground and are consequently impossible to see as they move about can also be followed by this method. The animal is trapped once, and need never be trapped again after a ring has been fitted around its tail. This ring contains a radioactive isotope, and its presence can be detected by a radiation counter suspended above the tunnel and on the end of a long rod, so that the experimenter need be nowhere near the animal. The radiation is heard as a series of clicks through headphones. In this way, the animal's underground movements can be followed and plotted, and when it opens up new tunnels this can be easily detected. Tagging a pregnant female in this way will soon allow one to locate the position of the nest containing young.

Nevertheless, despite the existence of these methods, the natural behaviour of animals in the field is frequently studied *en masse*. There are two reasons for this: firstly there is a genuine interest in the interaction of many individuals of the same species when grouped together – social behaviour, as it is called – and secondly it is obviously more economical of time and effort to have many creatures grouped together in the same place. The pursuit and analysis of the behaviour of solitary animals clearly present greater problems, as we have seen. In field observations of natural behaviour, therefore, it will be found that studies of group behaviour predominate. But much can be learned from the use of tagging methods of studying the behaviour of animals in the field, as the following examples show.

A good example of field study which of necessity must be carried out under particularly adverse conditions is that of penguins. Penguins are among the orders of birds that have lost the power of flight during evolution, and have taken to life in the sea. They swim by means of wings which have become modified as flippers, though the feet are webbed. They are found only in the southern hemisphere, mostly the Antarctic, and many species live among the ice floes, obtaining their food from the sea. In summer, however, they come to land to breed and raise their young, and this provides an opportunity for observation, albeit an arduous one. The usual methods of field observation are employed – the use of the hide, the marking of individuals, and photography for later analysis – and in conditions which may include gales of sixty miles per hour and temperatures of twenty degrees below freezing – even in the so-called 'summer'.

But the breeding colonies are vast, containing many thousands of individuals, and so an observation tent set up almost anywhere in the breeding area will enable close observation of many pairs. A technical difficulty arises in that the short legs of these birds makes the use of the usual leg-bands unsatisfactory, so that flipper bands have to be used instead. These are arranged so that the number printed on them can be read at a distance by the observer through binoculars. In penguins there is no sexual dimorphism so that both the sexes look alike to the observer and can only be established unequivocally by killing the individual whose behaviour has previously been observed, and dissecting it to reveal whether its interior sex organs are male or female – that is, testes or ovaries. This procedure is not so savage as it may sound, since the birds are very numerous, as has been noted, and large numbers are normally killed to provide food for men and sled dogs at near-by scientific or other stations.

Yet it is clear from the details of many patient hours of observing that the penguins themselves recognize each other's sex and what is more recognize each other as individuals, recognize and feed their own chicks, and can find their way back to the same nesting site, year after year, even though it may be buried on occasion under several feet of snow.

The male comes ashore from the sea by leaping up to the edge of the ice from a surface dive, proceeds to the colony site, and takes up position at a nesting site, perhaps the very one he bred at last year. Here he is typically joined by a female, whom he attracts by a display of the sort which is often found in birds. In the Adelie penguins it consists of stretching the neck and pointing the bill skywards and at the same time flapping the flippers and uttering a noisy cry. Once pair formation has taken place in this way, copulation follows, the male treading the female as is usual in birds. At this time the nest is built, though this is usually a poor affair consisting essentially of a pile of small stones, and indeed some kinds of penguins make no nest at all and incubate the eggs on the feet. During all this time both birds are fasting, since there is no food for them on shore – apart from the snow, at which they sometimes nibble. As soon as the two eggs are laid, the female goes back to sea for about two weeks, and the male carries on the incubation, only to be relieved by the female, who stays until a few days before the eggs hatch, when she has another shorter spell away. At each of these change-overs, the returning mate will proceed, more or less unerringly, from the sea to the appropriate nest in which its mate is lying, in the midst of many hundreds of others, greet the mate with a special form of display, and be greeted by the mate in turn. The oncoming bird will then take over the incubation and the one thus relieved will stand around for a while, often adding a stone or two to the nest heap before going off in its turn to the sea to break its fast.

When the eggs are hatched this routine goes on, though the visits to the sea alternate with greater frequency, for now the chicks themselves must also be fed. This is done by regurgitating swallowed food, the chick putting its beak into the throat of the parent, who stands above it with its mouth open and pointing downwards. When the chicks are about four weeks old the parents begin to stay away at the sea feeding for longer times, and both together. The chicks, thus left alone, start to leave the nest site and to gather in groups (termed 'crèches' – a poor name since it implies the presence of adult supervision, which is not seen). This is probably a protective mechanism against the predation of the skua bird, which is the only dangerous enemy of the penguin on land. The parents return from the sea from time to time to feed their young, however. They recognize their own from among the several hundred which may be grouped in the crèche, since they will yield only to their importunings for food, and not to those of other chicks, whom they will peck at and chase away. As the chicks get older and bolder they chase their departing parents, begging for more food, and incidentally learning the way to the sea in the process. By the eighth week the crèches are less in evidence: the young birds now tend to occupy the original nest sites. Soon they start moving to the edge of the sea and shortly after themselves take to the water to feed. They share the apparent apprehension shown by the adult birds in actually entering the water – they jostle for who's *not* to be first. This behaviour is in fact justified, since leopard seals – thought to be the most important predator of penguins in the water – are known to lurk beneath the edge of the ice floe from which the birds jump!

All these facts have been established by the field study methods of marking birds, from patient observation of them over many hours throughout many months and years, combined with the exposure of many feet of ciné film and other photographs and the interpretation and analysis of these records later. Not all problems relating to the behaviour of these fascinating birds have thereby been solved, of course. As yet, it is not known *how* they achieve the apparently difficult feat of recognizing each other – they all look identical to us – but it is of great importance to establish first that they do so, before attempts are made to devise ways of discovering how they do it. All science is strewn with examples of people offering ingenious explanations, and often ingenious experiments to support them, of alleged 'facts' which later work has shown to be untrue.

Seals, sea lions, and walruses are another group of animals which live at sea and come to land on ice floes to breed in large colonies. They are among the groups of mammals which have returned to the sea during the course of evolution – other examples being the sea otter and whales, dolphins, and porpoises. They live in the sea and, like the penguins, return to breed on the coast – land or ice floes – in large colonies. In consequence their social behaviour has been studied in some detail. Unlike the penguins they are in general polygamous – one male having many mates, the ratio of males to females in the harems of sea lions and fur seals being as high as one to forty. This results in large bachelor groups of unmated males in the breeding colonies, and the dominant males or harem bulls maintain a clear path around their harem – a sort of 'no-man's-land', or rather 'no-seal's-land' – and attack any strange male who dares to enter it.

Another species whose behaviour in the wild has been relatively well documented is the wolf. In Alaska the timber wolf was thought to be an important predator of a species of wild sheep which it was desired to preserve. A study of these wolves, partly through the analysis of the contents of their droppings, showed that they were not so culpable as had been feared. They did eat sheep, certainly, but frequently only the old or injured succumbed to their attacks, and the hale and hearty were more skilful and sufficiently agile and fleet of foot to escape them. Their diet also consisted largely of small animals like ground squirrels and mice. The composition of various packs was studied in detail and shown to consist on occasion of more than one family, so that the wolf cubs were all reared together in the same den. They were quite friendly among themselves, but hostile to strange wolves not belonging to the same pack.

Wolves hunt at night and rest at the den during the day. They travel long distances in search of prey. These naturalistic observations are of great interest in view of the domestication of the wolf into the various breeds of dogs. Wolves do well in captivity and cross-breed readily with dogs. The theory that domestic breeds of dogs are derived from two sources, the wolves and the jackals, is based on certain differences in their behaviour and is a plausible one; but the weight of the evidence from other characteristics seems to suggest that the wolves alone are responsible for all our dog breeds.

Then there is the special case of field observations in an enclosed or confined area so large relative to the size of the animals that it can essentially be regarded as a piece of their natural setting, but also affords special possibilities for observation and sometimes even experiment.

Aquaria are good examples of this type of arrangement, and from observation in aquaria much has been learned about the behaviour of underwater creatures. The enormous aquaria found in several places in America provide, as well as amusement for visitors, fine opportunities for observing the behaviour of larger marine life, and of these the dolphins are among the most fascinating. Here is yet another example, like seals and sea lions and whales, of mammals that have returned to the sea in the course of evolution. But, like whales, and unlike seals and sea lions, their return has been quite complete, since they are no longer dependent upon returning to a shore or ice floe to breed and give birth to their young. Every phase of the mating and parturition of dolphins has been studied by means of underwater observation windows in the sides of the tanks, from the initial pre-copulatory or courtship activities of the pair to the care of the young after birth. It is difficult not to think of dolphins as fish, in view of their shark-like shape, but there are extremely important differences: apart from anything else, they are air-breathing mammals which would drown if they were prevented from surfacing to breath from time to time. And one of the most striking differences is in their mating and maternal behaviour. In the dolphin, prolonged companionship of male and female in the spring is followed by mating. Like all mammals the male of the species possesses a penis which must be inserted into the vagina of the female in order to ensure the ejaculation of semen and fertilization. Now this organ is usually hidden within the genital slit, but during sexual excitement it becomes erect and springs out quickly. Copulation takes place when the female rolls on to her side to allow the male, also swimming on one side, usually the right, to penetrate her. Sometimes the male will approach from underneath, swimming on his back.

The female gives birth about one year later and the infant dolphin is usually born tail first. The mother is often attended by another female who swims alongside meanwhile. The baby keeps close to the mother, who allows it to suckle from the mammary glands located on either side of the genital opening. The milk can be ejected by the mother, and it seems that the stimulation of the nipple is not necessary for this purpose. If the youngster does not keep near the mother, she swims after it in order to get it into line, literally, since it means positioning it close to her side. If the young persistently evades her, she may turn over on her back and surface, with the baby caught on her chest between her two flippers, thus lifting it out of the water and effectively immobilizing it for a time. When older, the young are allowed to stray further away, but they continue to be suckled by the mother until they begin a gradual weaning process at about six months of age. Now it is clear that without special facility for observation it would have been practically impossible to know so much detail about the life history of these fascinating creatures.

Another example of the device of taking the laboratory into the field is the use of the terrarium, or large open sink, usually made of concrete, within which the natural conditions of the species being studied can be simulated while denying them the opportunity of wandering away from the investigator at any time. Such a scheme was used for the study of social behaviour in crabs of a species which normally live in tidal mud banks. It therefore became necessary to construct a mud bank of the right sort of mud, heavily impregnated as it was with salt water, and then to flood the crabbery each day with water from a garden hose, and draining it later so as to create the illusion of the rising and the falling of tides. This stratagem was very successful, and the crabs flourished. The only problem was keeping the frogs out of this desirable residence, but a wire fence solved that! The observer sat on a platform near the pool with a spy-glass and could see everything that went on. Marking individuals with paint enabled the experimenter to keep trace of the progress of fights, courtship between pairs, and so forth in a way which would not have been possible on an open beach. The use of the crabbery confirmed a previous finding that the species of crabs known as fiddler crabs, because of the disproportionately large development and conspicuous colouring of either one of the two front claws, show a strong tidal rhythm. This is seen in the way their phases of activity superseded each other. These included simply feeding and digging in the mud, and wandering around, either in a simple exploratory way, or in an aggressive way, challenging and fighting other crabs met with, and ejecting them and driving them away from their burrows in the mud which the wanderer then occupies temporarily. In this way one individual comes to be boss animal to which all others give way, with a number two crab which retreats before the attack or threats of number one only, but dominates all others, a number three who submits to one and two only and so on. But in crabs it was found that these dominance hierarchies, as they are called, lasted for a few days at most, in contrast to the more permanent ones seen in other species which live together in groups. Then the crab settles down to defend a particular territory around its own burrow and behaves in a way characteristic of these crabs. The large claw is waved in the air in a roughly circular motion, rather slowly; this is called a display. Mating may occur in this phase, and if a female approaches the male near its burrow, the male displays differently by tilting its shell backwards and waving its great claw higher in a wider arc and faster than before. The female may then descend the male's burrow, where copulation presumably takes place, though pairing may also be observed on the surface.

What then are the disadvantages of studying animal behaviour in the field? The advantages we have seen – they can be summed up in the phrase 'naturalness of conditions of observation'; the disadvantages stem from this same condition. In any scientific observation the analysis of what causes what is crucial. In animal behaviour we are usually concerned with what changes or conditions in the animal's environment – the different stimuli as they are called technically – give rise to what alterations or modifications in the way it acts – the different behavioural responses, to use this technical term again. Now in the field it may be possible to observe the responses the animal makes quite satisfactorily, but it is rarely possible to know exactly what caused them, in other words to identify the stimuli impinging on the senses – sight, sound, smell, touch, etc. – and to evaluate their relative importance. Moreover, it is usually impossible to control the nature of the stimulus in the way in which it can readily be controlled inside the laboratory in order to analyse the relative importance of its component parts. The sort of thing one is faced with in nature is the effect upon the animal of some complex stimulus, like the arrival of another animal of the same or different species, or a seasonal change such as an increase in the length of day, or a large and perhaps catastrophic change such as a decrease in the food supply. Now in science one usually proceeds from the understanding of the simple to the understanding of the complex, and in order to find out how the complex operates it is often necessary to reduce it to its simplest elements and to study them in turn, one by one – or at least to identify the elements so that their effects may be distinguished separately. For example, what is it that makes male birds sing in the spring? Is it the presence of females, or the increase in length of day, or possibly the increase in temperature? In the laboratory it is possible to change each of these components individually in order to assess the effect upon bird song, but in the field it would be very difficult to do so. However, it is sometimes possible to attempt some of the sort of control which seems desirable over those aspects of stimuli which seem to be important in determining the response. This involves modifying the natural conditions of the animal to a greater or smaller extent, but there are few situations in which it is possible to do this quite radically without sacrificing the advantages of the field situation. If we cannot bring the animal into the laboratory, let us take the laboratory to the animal; and much ingenuity has been shown by experimenters in devising ways of doing this. For example, it is possible to follow the animal into its natural setting with some of the apparatus of the laboratory and then to try to use it on the spot. This technique has been employed in the study of small mammals in American forests, especially the chipmunk and the ground squirrel. These species are ground-loving, and so the complication of vertical movement was minimized, though in one case a vertical version of the standard laboratory maze (to be described in the next chapter) was pressed into service. Individual squirrels and chipmunks were trapped and marked for subsequent identification and then lured by food into apparatus where it was hoped that they would learn problems involving the discrimination of symbols on different doors – the one being designated correct leading to further food. But one squirrel was observed for 200 trials spread over seven days in such an apparatus with no evidence of improvement due to learning! Another technique was to trap the animal into the apparatus, usually a maze, and then allow it to find its way out in order to regain freedom. In two cases it was possible to identify the same animals the following year and then to test them again. If anything, they did less well the second time, suggesting perhaps that they had been affected by their long winter sleep during hibernation. These methods have not been widely used and it is not difficult to see why. The sacrifice of the important aspects of control over the individual subject and its background, possible in the laboratory, is made with very little compensation in 'naturalness' of surroundings in the field. The apparatus itself inevitably introduces an un-natural element into the field situation which might be held to be objectionable.

More successful in the search to modify environmental stimuli in the natural setting have been those attempts which have concentrated on only one aspect. An example is the study to determine which qualities of the gull's egg induce the brooding behaviour. Is it the size, the shape, or the colour and markings? Large eggs, small eggs, plain eggs, heavily spotted eggs, and eggs of strange shapes were substituted in gulls' nests in the field and the effects observed. It turned out that whereas colour, pattern, and size are important, the shape of the egg does not matter, though it can be shown independently that the gull can distinguish the various shapes used. Thus an egg that is $1\frac{1}{2}$ times the normal in size and has lots of contrasting speckles will stimulate a more intense incubating behaviour from the bird than the real thing. Not only does it act as a normal stimulus, it is a supernormal one! So it can be seen that the scientific observer of behaviour can often gain some control over those aspects of the environment which he suspects affect the behaviour he is interested in, and he can proceed to alter them in the field situation.

However, the whole position is vastly more complicated than might be inferred from the simple considerations of stimulus and response mentioned above. Behaviour does not result from a simple stimulus, in the way the insertion of a coin causes the appearance of a stamp in a stamp machine – quite automatically. Rather it is like the more modern elaborate food-vending machines, which need a source of power to keep them going and deliver some foods hot and some refrigerated. To be sure, putting the coin in starts something going, things go round inside and the item selected is delivered, sometimes on a moving belt. But cut off the power and all this stops. No amount of stimulation in the form of coins will now deliver any response in the form of goods. Though this also is too crude an analogy for what happens when an animal behaves, it is a shade less crude than a simple stimulus-response idea as exemplified by the stamp machine. There are important processes going on in the animal before the stimulus occurs and between the stimulus and the response, and it is these that are sometimes crucial in our study. Let us say that we know that an animal responds in a certain way to a certain stimulus and we know the characteristics of the stimulus which are important. But what is the part that, say, hunger plays in the way this process operates? Does the response only occur when the animal is hungry or would it operate during satiation also? Now this is the sort of question which it may be impossible to answer from field observation. Granted that it may be possible to obtain precise definitions of both the stimulus and the response in this circumstance, it might still be very difficult to know whether or not the animals being observed were hungry and, if so, how hungry. If the animals are in captivity one has as complete control over each and every aspect of their previous existence as may be desired. Conditions antecedent to the experiment can be varied beforehand so that the animals come to the experiment short of food, satiated, having been kept in the dark or in the light, in the heat or the cold, or what you will. If the animal is free to roam, then it is never possible to be entirely sure in such respects.

Then there are the questions which can be answered only by modifying the state of the animal in some way rather more drastic than the relatively mild changes required to alter its feeding habits, living conditions, and so on as mentioned above. Sometimes the need arises to alter the nervous system in some way, often by surgically removing part of it, especially from the brain, or by administering some drug which affects its functioning. Such procedures can clearly only be carried out successfully in the laboratory, and special precautions need to be taken to ensure success of observations made under such conditions.

3

The Study of Animal Behaviour in the Laboratory

WE have seen how the behavioural scientist goes about the study of animal behaviour in the wild. How then does he go about the study of animal behaviour in captivity? First, as the White Queen said, the fish must be caught. By which is meant that the first requisite is a suitable collection of animal subjects. Such a collection may be found in a zoo, and some notable research work on animal behaviour has been done in zoological gardens; but in general the animals used for laboratory experimentation are maintained in colonies specially organized for the purpose. Such colonies are either businesses run on a commercial basis which supply animals to laboratories to order, or they may be maintained by the laboratories themselves. The latter is often the case with psychological studies, since the way the animal has been reared and the influences which have acted upon it in the period before experimentation begins may often be of importance in determining the results observed. Such control is also crucial in genetical experimentation, where control over breeding is absolutely necessary so that the heredity of any individual animal can be established without any doubt by reference to its parentage. But in other biological sciences there is not the same emphasis on the importance of the pre-experimental experience of the animals used as subjects: given that they are of a certain age and weight it may not be necessary to know more about their background than that. But the psychologist, being concerned about what has happened to the animal prior to the experiment, often uses only animals born within the laboratory. This enables a control to be exercised over all the details of husbandry – not only the diet fed to the animals and their general health and the temperature of the colony room, but also more subtle influences, largely psychological in nature, such as the size of the cage, the number of animals kept in each cage, and the number of times the occupants are handled by humans during the course of feeding, watering, and cleaning. It may not always be possible to assess exactly the effect of variations in such things, but there is some evidence now that some of them can affect the outcome of certain types of experimentation; and while we are still unable to be sure of which does what, many investigators play safe by ensuring so far as possible that all the animals used in a particular experiment have all had the same previous experience, both in terms of the way they were reared and in their previous participation in other experimentation. This is done in the hope that any effects of such experiences will be the same for all animals and can therefore be discounted as causal factors in the way the animals behave in the experiments subsequently carried out on them. The similarity of background experience, together with a careful choice of the hereditary characters which the various strains are known to possess, will result in a group of animals each of which resembles every other one very closely. This uniformity of the subjects in an experiment is of considerable importance to the laboratory experimenter. In the field, on the other hand, the behavioural scientist may have to content himself with observations on a few animals, because of the difficulties of locating suitable subjects for study – and here we are not speaking of assessments of the behaviour of large groups as a whole, with little emphasis on the behaviour of individuals constituting the groups, as in flocks of birds. The scientist will, therefore, have to make allowances for the fact that those animals he does succeed in observing may have been subjected to different prior experiences and may also have differences in inherited capacities which again could affect the behaviour being studied, so that the part or sample may not be truly representative of the whole. Thus the conclusions the scientist draws will be more tentative for a given number of subjects in the field than they would be for the same number of subjects studied in the laboratory, where the possibly disturbing factors due to heredity and previous environment are controlled in the ways described. To put it technically, the less variability of response encountered the smaller the sample you need to study.

To take a hypothetical example, if you want to measure the length of time it takes cats after they have come into a room with an open fire to curl up and go to sleep on the mat in front of it, you would, in the laboratory, always use the same room and the same mat, adjust the fire so that it always gave off the same amount of heat, and so on. That is, you would standardize the stimulus situation, then you would choose cats of the same age, with the same previous experience of open fires, and all from the same strain, etc. The latter would ensure that they all had the same colour of fur, which you might think important in considering the amount of heat absorbed – black being better than, say, white in this respect. Then you would perform the experiment at the same time of day, the same number of hours after the cats had fed, and the same number of minutes after keeping them in an outer room at a fixed, perhaps rather cool, temperature, and so on. Indeed there are probably many other factors which you might want to control and which you could standardize or make allowances for in the laboratory. The field worker, however, is rather in the position you would be in if, to pursue our hypothetical example, you could only time the cats going to sleep in their own homes, in their 'natural setting', and you had to visit each in order to do so. Now, as a result, in collecting your sample all sorts of things would vary, the size of room, the type of fire, perhaps the time of the visit, to say nothing of the type of cat you found, its age, sex, etc.

Now in computing an average time taken to go to sleep you would get a lot of variation due to these factors, so that in order to get a result about cats' sleeping habits which you were reasonably sure approximated to the truth you would need a far larger number of cats whose sleeping times you had taken than you would if you had been able to make the controlled observations in the laboratory in the way described. But, it might be argued, the average so derived from the home visits would be a general statement about cats, and hence of more practical use, covering as it does a wider range of situations, than the laboratory study, in which you might have deliberately restricted yourself, for the purposes of the controls desired, to, say, one breed of cat, one fixed time since the last meal, one fire temperature, one mat thickness, etc. And so it would be, if this were really the way experimental work was done in the behavioural laboratory these days. But it is not, and still with the same number of cats as had been visited in their homes, or even fewer, and without a greatly increased expenditure of effort in the laboratory, observations could be conducted in which several of the different conditions mentioned were systematically *varied* while the rest were held constant. Thus we might use two different breeds of cats, three different times since feeding, four different thicknesses of mat, etc., while making sure that all the cats were the same age, had had the same size of meal beforehand, and so on. In this way, then – and here is the important advantage of the controls possible in the laboratory – we should have, not only the same general average as we would have obtained from the visits to the cats' homes, but also we would know the effect of each of the items we had varied, both singly and in combination with each other. Thus it might be that time since the last meal had no effect on one breed of cat, whereas in the other the shorter the length of time the faster they went to sleep – whereas thickness of mat had no effect upon either.

This example of the way the behavioural scientist in the laboratory might 'dissect', as it were, the particular piece of behaviour of cats known as 'going to sleep' into the various factors or variables responsible for it might quite properly be held to be unrepresentative of the sort of thing we might want to observe in an animal in the field in any case. And it would also be a reasonable criticism that the advantage of laboratory controls over both the prior experience of the animals and over the stimulus conditions which they are to experience is not always so apparent as in this fictitious example. There is usually a complexity of factors relating to the organism itself and its behaviour which have to be taken into account in deciding whether or not they should be taken into the laboratory for the purpose.

By far the largest laboratory colonies of experimental animals for behavioural research are composed of rats, though mice are often used as well, and occasionally other small mammals, hamsters, rabbits, and racoons. Colonies of primates – monkeys and apes – are also found, but their use is less widespread, partly for reasons of expense. The rat used is almost always the white or albino rat. This is a mutation from the brown or Norway rat and has become domesticated in the service of man as a laboratory animal since it was first put to this use some hundred years ago. In its wild state the Norway rat is the common brown or grey rat of the farm and the cellar throughout Europe. The black rat, or ship's rat, found in some dock areas, is a different species and is not used in laboratory work. Albino rats are docile and friendly and have white fur and pink eyes. They are, after man himself, one of the organisms most subjected to scientific study and this is probably especially true of their behaviour. Some of the first laboratory experimentation upon animal behaviour was done on the white rat – appropriately enough it was a study involving the use of a labyrinth or maze, which afterwards became and remained for a long time the standard instrument for the study of learning and reasoning.

Animals in a colony are almost invariably kept in cages, which, contrary to what is often thought, are designed primarily as places for them to live in, not as a means of confining them in captivity. It is a misapprehension to assume that laboratory animals are rabidly trying to break out of their cages all the time. It is striking how often rats whose cage has been inadvertently left open overnight will wander around the laboratory but later return to it and are to be found there the following morning quietly sleeping off the effects of a night out! Striking, but perhaps not surprising when one reflects that the home cage is, after all, the place where the animal receives its major satisfactions in life – food, water, sexual experience (if it is used for breeding), and so on. The life of a laboratory rat can be envisaged as a life of leisured tranquillity, perhaps interspersed with some periods of concern about when the next meal is coming or some moments of fairly acute discomfort. Every effort to ensure the animal's comfort, well-being, and tranquillity that ingenuity and assiduous and careful attention can devise is made. Indeed, so otiose an existence do they lead that proponents of the observation of wild animals in the field claim that the organism studied by the laboratory worker is an unnatural and artificial product of his man-made environment and that consequently the value of observations made upon such material is limited. Their premise is correct, but their conclusion fails. This is the price we must pay to secure that scientific rigour in the study of certain aspects of behaviour which is only secured within the laboratory. And the additional difficulty in interpretation thereby created is not as large as might be thought – after all, civilized man is no longer a wild creature himself, but an artificial product of his man-made environment. Another misapprehension among visitors to an animal laboratory relates to the size of cage used. It is often thought to be too small, but with nocturnal animals such as the rat or mouse, which spend most of the day asleep – since it is their 'night' – the size of the cage is immaterial after a certain point. However large, the animals will aggregate, that is get together and sleep in a heap. At night, when they become active, they will race around the cage and display violent activity even in a relatively small space. With larger animals the problems are greater, and especially with monkeys and apes there is a need to provide some amusement. This is often done by means of playthings, ropes and swings, etc., of the kind frequently used for the same purpose in zoos.

The health and welfare of the animal colony is thus a source of constant concern to the behavioural scientist. The kind of diet fed to the animals is carefully considered and expertly made up. It is now possible to rear and breed many small laboratory animals solely on a diet of tap water and compressed food, not unlike dog biscuits or cattle cake. Some laboratories have automatic devices for delivering this food and water to the animals in their cages; sanitary arrangements may also be automatic. The animals have suitably sized wire-mesh floors to their cages so that droppings and urine pass straight through and are washed away from time to time. Alternatively they may pass into a tray filled with sawdust (or a deodorant chemical which acts in the same way), and which is changed frequently. Usually animals of like sex are housed together in small groups. Animals are not usually housed singly, except for larger and consequently more valuable animals if there is any danger of their hurting each other if fighting should occur, or for experimental purposes. The sexes are not usually mixed except for breeding, and then the male is usually removed after conception has taken place and before the young are born. Care of the young is usually left to the mother, except in special circumstances, as for example in the case of a valuable ape baby whose mother is dead or who rejects it, or for experimental purposes. Experiments might include the study of infant reactions to various kinds of substitute mothers, or delivering the offspring of small animals such as rats or guinea pigs directly into a sterile, aseptic environment in order to start a strain of germ-free animals.

And so the laboratory animal is raised in these optimum conditions until it is needed for experimental purposes. Many animals in a colony will not serve as subjects for experimentation at any time during their lives: they will be needed for breeding purposes to maintain the supply in subsequent generations; others will not be used because they do not conform to experimental requirements of age or size or some other feature of importance to the experimentation going on at the time.

What kinds of behavioural experiments take place in the laboratory? As might be expected they tend to be concerned more with individual than group processes of the sort we have seen exemplified in the field studies discussed in Chapter 2, and with inquiries which by their very nature tend to need more or less elaborate arrangements in the way of instrumentation. This is not to say that many animals may not be involved: on the contrary it might need a lot of work on a large number of animals to establish, for example, that cats are essentially colour blind. What usually happens is that each animal is studied individually in the apparatus and the results are recorded for it separately; then the whole set is analysed carefully by those special statistical techniques which are appropriate for establishing the presence or absence of the phenomenon – in this case, colour vision.

A special problem of laboratory experimentation lies in the motivation of behaviour – or to put it crudely in getting the animal to behave in the way you want. In the field this is rarely a problem: you observe what the animal is doing – its motives for doing so may be the object of study or of speculation – but only rarely is there any question of experimental intervention with those motives. The same is not true of laboratory work, and indeed a major division may be drawn between those types of laboratory investigation which employ some artificial system of motivating the animal and those which, like much of field experimentation, do not need this. The word 'artificial' is used here not in the sense of unnatural but rather in the sense of contrived, and certainly the motivation is external to the animal, not a spontaneous reaction. Let us consider first laboratory work in which no external motivation of this kind is applied.

And here we must make another distinction, this time between the short-term or acute experiment and the long-term or chronic one. Obviously any such distinction is bound to be somewhat arbitrary: few things, especially in science, are clear cut, black or white. They are usually a matter of degree, and this distinction is no exception. However, it is convenient to distinguish between those behavioural experiments which involve short periods of observation, sometimes accomplished in one or two days, and those which require repeated observations over many days, weeks, even months, and sometimes years.

Observation of simple cage behaviour of rodents might serve as an example of the short-term type of experiment which needs no special motivational techniques. Time-sampling techniques are often used, that is, each rat's doings at a particular time of day are noted and the findings from a relatively large sample of animals tabulated to reveal a pattern in the sequences of behaviour of the animals over the whole day – that is, to discover what sort of activity the rat indulges in, and for how long, and what follows what, etc. In this way it has been shown that male rats spend a large part of their waking time grooming themselves – rather as cats do, even down to the face-washing movements, though male rats use both paws rather than one. A simple variant of this technique of observation is to remove the rat from its home cage for a fixed time each day and place it in a strange place, such as a new cage or a specially designed arena or open space, and observe the consequences. A surprising amount of information can be obtained in this way about its reactions to novelty, which may include fear, often seen as emotional elimination of urine and faecal pellets in this species, as well as a tendency, decreasing with time, to explore the situation. The tendencies thus measured can be related to other aspects of the animal's behaviour measured in other ways: for example, the promptness with which it learns reactions based on fear.

An example of experiments in this same class (of those not requiring anything other than intrinsic motivation) but which can be classified as a chronic or long-term experiment is that in which the behaviour of the animal has also been studied over a long period in its living quarters, except that in this case those living quarters have been modified for experimental purposes. The study of the voluntary exercise of the rat and mouse exemplify this approach. Here the animal lives in a special cage which has direct access to an activity wheel. This is a treadmill of the type often seen attached to the side of the mouse boxes found in pet shops. In the laboratory they are carefully calibrated to tell how far the animal has run each day. When the animal has settled down in the new cage, the level of its voluntary activity as measured by the wheel is a good indicator of its state of health and its responsiveness to other influences. Thus it has long been established that the high peaks of activity which occur every fourth day, to be found in the cage-activity records of female rats, are correlated with the menstrual cycle, which also has a four-day rhythm in this species. The change in activity which follows the administration of various drugs can be observed by this method and is used to study the resemblances between the effects of various compounds.

It may be thought that cage activity as such is so gross a piece of behaviour that it hardly merits detailed attention; yet in animals as low in the philogenetic scale as the rat and mouse, it is not always easy to distinguish aspects of behaviour which can conveniently be studied. After all, all activity is behaviour – stop behaving, you might say, and you die – and in the absence of more complex activity general body movements are of great interest to the behavioural scientist.

But lest it be thought that this is too easy, some of the difficulties involved in even this type of apparently simple measurement should be mentioned. Firstly there are the problems associated with the measuring apparatus itself, and secondly there are those associated with the way the measurements are made. Let us consider the apparatus first. There are several kinds of apparatus currently used for measuring cage activity of small animals. Most of them, like the treadmill described earlier, depend for their operation on converting the movements the animals make into mechanical movement of some part of the apparatus. Thus, as the rat steps on the treadmill, which is free to turn, it moves in the opposite direction, and when one complete revolution of the wheel is completed, this activates a counter. Alternatively the cage may be mounted on tambours full of air, so that as the animal moves around in it the volume of the tambours changes and the resultant change in air pressure is communicated through a system of rubber tubes to a pen, which consequently alters its position on a moving strip of paper, thus graphing the animal's movements continuously. There are many variations of these methods, but they all have intrinsic difficulties arising from the fact that the inertia of movement almost inevitably varies from cage to cage – that is, it takes a different strength of movement in each case to operate the mechanism. Combined with this is the difficulty that animals differ in weight, so that they themselves generate different mechanical forces with the same apparent movement. These difficulties are particularly apparent when it is desired to use this method for measuring the relatively slight and subtle changes to be expected from some experimental treatment. Thus it may be necessary to rotate the animals around the available cages, so that any differences between cages in the way they transmit the activity can be spread over all the animals involved, and will not become associated with any one animal or group of animals and so, perhaps, be mistaken for a peculiarity of these animals, and in turn for the result of a particular experimental treatment they may have been given. This is how error can creep into otherwise carefully thought out experimental work, though when detected it is usually dignified by the name of 'artefact'. But changing the measuring cages around in this way also creates its own problems, because the animals may be disturbed by the change to an unfamiliar cage and this will in turn affect the results, probably by reducing activity. Allowance may therefore have to be made for this. The changes associated with the different phases of the menstrual cycle in female rats mentioned earlier are, however, so large that they are easily detected with the crudest mechanical apparatus. For the detection of smaller changes it becomes necessary to use more sensitive apparatus without defects of the mechanical sort referred to. Usually this means that some form of electronic device is used, and of these the most widely employed is probably the photocell. This is a form of radio valve which is sensitive to light in the way that a photographic light meter is. When a beam of light is shone upon it, it can be arranged to hold open an electrical circuit of which it forms a part. But break that beam of light, and the decrease in the intensity of the light arriving at the cell increases its resistance sufficiently to cause it to reverse and close the circuit. The number of times this happens can be recorded by a counter, and it can easily be seen that if such a system is arranged so that the animal, in its movement around its cage, interrupts the beam of light as it passes, then we have a system of measuring its activity which is independent of any mechanical device.

But, as before, this solution creates its own problems. Firstly, it may not be desirable to expose the animals always to illumination – it may be necessary to give them a period of darkness in the usual way. A low level of illumination, especially a red light, may obviate this difficulty. Secondly, in order to activate the circuit, the animal needs to be actually moving around from place to place within its cage. Now, it may be that quite a lot of activity can be indulged in which does not result in such movement. Thus, a rat might sit in one spot and scratch itself furiously without preventing the light from reaching the cell. Nevertheless in doing so it would create enough mechanical movement to operate a more conventional type of mechanical recorder.

The problems involved in this apparently simple laboratory measure of cage activity have been stressed in order to point out the dependence of the behavioural result obtained upon the type of apparatus used to measure it, and the need for caution in interpreting the findings from experiments which rely on a particular piece of experimental apparatus for their outcome. The cautious experimenter is constantly alert for the sort of artefacts we have discussed.

Just as some experimenters, as we have seen, have taken the apparatus of the laboratory out into the field and used it with some success there, so others have sought to bring the field into the laboratory, so to speak, and to combine the advantages of both situations. The most ambitious of these attempts has been the prolonged study of the effects of size of population upon the behaviour and pathology of rats. It was first found that wild rats left undisturbed with a superabundant supply of food and water in an outside enclosure showed a self-limitation of size of population. Thus in a large area enclosing a quarter of an acre the population – descendants of five original females – stabilized at around 150 animals after some two years, whereas there was room physically for many more, and indeed the rate of reproduction should have soon led to a population of around 5,000. Rats are very prolific!

In order to study what was happening somewhat more closely, a simulated natural environment was set up in a barn. Typically four interconnecting pens were used, with artificial burrows containing nesting boxes and a generous supply of food, water, and nesting material. Arrangements were made to control the size of the initial populations, and careful observations were kept on various activities – fighting, mating, caring for the young, and so on. It was found that there was a definite tendency for the animals all to crowd into one of the pens for feeding, even though each was provided with food. And it was this crowding together that was the cause of the failure of the population living in these apparently ideal conditions to expand. Pregnant mothers, and especially nursing mothers with young, were disturbed not only by the constant traffic but also by molestation, and this caused them to abandon and even to eat their own litters. Indeed, in some of the pens, where one buck rat had established an ascendancy over all other males and quietly ruled a 'harem', the breeding performance was superior. It seems that there may be an optimum size for a rat community – probably around twelve – and that increases over that size give rise to social pressures which are disruptive of normal family life. In this way there may be a natural limitation on the possible expansion of a population, even when there is not necessarily any severe competition for food, such as is undoubtedly the case in the population fluctuations among the lemmings to be discussed in Chapter 6.

So far we have been concerned with the type of laboratory experiment in which there was no special need for arousing the motivation of the animal subjects. Now, in turning to the type of experiment in which it is necessary to do this in order to elicit the particular behaviour it is desired to study, we encounter a bewildering variety of possible approaches and techniques. Here again let us first distinguish between the short and the long term, and as an example of the short-term or acute experiment let us consider the procedure known as 'escape-avoidance conditioning'. Imagine a metal box about the size of a domestic sink but divided across the middle by a partition which has an opening in it. The walls are made of smooth metal sheeting, but the floor is composed of two grids made of metal rods, one grid on either side of the partition. Each of these grids can be electrified so that a rat standing on it will receive a slightly painful, though in no way damaging, shock to the feet. This will give rise to sensations of pain, and in turn fear and a desire to run away, and this constitutes the motivation upon which the success of the experiment depends. This type of acute motivation, aroused in this or some other way actually during the course of the experiment, is widely employed and is very useful. In the present instance its object is to induce the animal to make a particular type of response, which the apparatus provides for, that is, to move through the opening in the partition to the compartment on the other side, and thus escape the shock. This is how the 'escape' part of the name arose. But even this is not as simple as might at first be thought: laboratory animals do not necessarily do what you expect of them, and it was some while before the technique was perfected. For example, if the walls are not smooth the rat may escape, not by running through the opening to the next compartment as the experimenter intends it should, but by jumping up off the painful grid and clinging to a projection such as a ledge or seam in the metal or even a large screwhead. If the bars of the grid are too far apart it may escape by squeezing through them; if they are themselves too broad it may escape by balancing on one bar only, thus avoiding completing the circuit between adjacent bars which form the electrical positive and negative necessary for producing the shock. Rats have even been known to learn that shock cannot reach them if they straddle alternate bars!

Assuming, then, that these possibilities are excluded, we now have a reliable method of inducing a rat to move from one compartment to another at will. What happens then? The current is switched off in the compartment where the animal was originally confined and switched on in the compartment into which the animal has moved, whereupon it will turn round and go back into the first compartment. In this way it can be induced to shuttle back and forth between the two sides of the box, which for this reason is sometimes called a 'shuttle' box. This then completes the escape part of the response; the avoidance is introduced in the following manner. If we flash a light in the box, or sound a buzzer in it a few seconds before the shock is turned on, this serves as a signal (a 'conditioned signal', as it is called) to the rat that the shock is coming, and it quickly learns to make the running response to get it to the other side *before* the shock occurs. Thus it *avoids* the shock, and this is how the avoidance part of the name of this technique arose. It is also termed 'conditioning' because the signal is in these circumstances technically a conditioned signal or stimulus in that the rat would normally be quite indifferent to it, and it has only gained the special property of occasioning the avoidance behaviour as a result of its association with the shock to the feet. That is, its effectiveness is *conditional* upon this association having taken place. Thus the whole procedure is also a learning process – the rat does not at once learn either the escape procedure of running to the other side of the box when it feels the shock to the feet, or the avoidance procedure of making the same move when the signal heralding the shock occurs. In practice, however, it takes only a few trials for this to happen, so that the experience for the animal is not particularly unpleasant. What has happened is that the animal has learned to associate the sound of the buzzer or the flashing of the light with the coming of the shock, and this learning process can be completed very quickly in as few as twenty or so trials in which the signal and the shock are presented together. This would take perhaps half an hour to accomplish, and thereafter the animal will always – with perhaps a few lapses from time to time – avoid the shock by crossing to the other side of the box as soon as the signal starts and before the shock comes on. To all intents and purposes, the shock could now be disconnected. Therefore, we now have not only a method of making the animal move from one compartment to the other at will, but we have a way of making it do so in response to a signal of our own choosing.

In this way we have gained control over the animal's behaviour in a manner which allows us to study the effect upon it of many other variables. Thus we can now study the relative effectiveness of different aspects of the situation within the conditioning procedure – the intensity of the shock, the effectiveness of the light signal as opposed to the buzzer, the effect of varying the intervals of time between the onset of the signal and the onset of the shock. All of these are of considerable technical and theoretical interest to the behavioural scientist and have been studied extensively. It has been shown that increasing the intensity of the shock increases the speed of learning to avoid it up to a certain point, after which it becomes less effective; that a buzzer is generally more effective than a flashing light; and that the best interval between signal and shock is around five seconds. But in addition to these 'parameters', as they are called, of escape-avoidance conditioning itself, we can use the method to tell us about other factors quite remote from it. For example, not all animals learn to avoid the shock equally well: some strains of rats are better at it than others, and this is a valuable clue to their responsiveness to the shock. The nature of the previous experience which the animals have undergone can also affect the way in which the avoidance conditioning proceeds, and this applies not only to events in the immediate past but also in the further distant past, for example during the animal's infancy. The influence of drugs, too, can also be studied in this way, by making comparisons between animals given a certain drug and others given another drug or a placebo, which is a dose of some inert substance without effect.

Turning to an example of a longer term type of experiment than escape-avoidance conditioning, which is accomplished in a single session without the animals having to be moved once out of the box, let us consider maze learning in rats and mice. This has already been mentioned as one of the oldest laboratory techniques in the repertoire of the behavioural scientist. Indeed, the popular image of the psychologist in the animal laboratory, at least in America, would be that of a figure in a white coat running a rat in a maze, and I suspect that the expression a 'rat race' used in a derogatory sense derives from the same source. Mazes as used in the laboratory may be of many types and designed in a variety of patterns with different degrees of complexity in the learning problem which they present. The earliest one whose use is noted was designed roughly according to the pattern of the famous maze in the grounds of Hampton Court Palace, near London. This maze is one of several in English gardens, and admission to it still costs only threepence. The passages in it are bounded by high, thick hedges. It is a relatively complex maze, but people easily learn to find their way to the centre, especially if they already know the route (left, right, right, left, left, and left again at the successive choice points). However, imagine a wooden maze of this pattern scaled down for the rat, so that the passages are some few inches wide and the walls correspondingly high, and roofed with wire netting to prevent the animal from escaping, and it will be seen that to learn the correct route from start to centre is quite a task to set before a rat. Indeed the tendency has been over the years to make the mazes used in animal learning much simpler, so that nowadays one frequently used pattern is the simple T-maze, in which the rat has to learn only one correct turn, that to the right or to the left after leaving the start. But what inducement is offered to the animal to attempt this task? Obviously with the human animal there are reasons for wanting to get to the centre of the maze, and these reasons may be very complex ones. 'Because it is there' is one which expresses the human response to any sort of challenge rather well, and there may be a multitude of other reasons which reflect other aspects of human striving, from the rather abstract level mentioned to what might be considered the somewhat baser motives of checking if the directions given above are correct (they are!) or even to getting one's money's worth. But with the laboratory animal none of these motivations operate, and some other way must be found. The usual procedure is to provide some reward for the animal at the centre or wherever the goal, as it is termed, is located. This is usually in the form of food or water, but in order to make this reward of any incentive value to the animal it is necessary to deprive it to some extent of food or of water before starting it out in the maze. This is done by withholding food or water for some hours beforehand, and indeed the usual procedure is to have a feeding schedule operating in the laboratory which takes account of the requirements of the current experimentation so that animals taking part in experimental work are fed their daily ration of food or water only *after* they have been run in the maze each day. This would be in contrast to the general colony arrangements as described earlier in which food and water are available to the animals in their cages twenty-four hours a day. Thus when the rat is taken from its cage and put to run in the maze it is probably quite hungry or thirsty, and so will eat or drink rather readily when it reaches the goal for the first time – indeed this process may have been deliberately accelerated by previously feeding it part at least of its day's ration in the goal box of the maze for several days previously in order to acclimatize it to feeding there. On reaching the goal the rat is allowed to eat or drink for a few moments, but certainly not long enough to satiate its appetite or completely to slake its thirst before it is removed and replaced at the start of the maze. Thereupon it finds its way to the goal again and has another nibble or sip. This process is repeated several times in succession and similarly on subsequent days, and soon the animal is running readily from start to goal, always making the correct turn or turns at choice points, and we can then say the rat has learned the maze. Here again we have gained control over a piece of behaviour by means of the experimental procedure we have adopted, and in its various modifications this maze-running procedure can provide a very sensitive technique for measuring variables which influence the learning process as seen in the speed the animal runs, the way in which running up blind alleys is eliminated, etc. And, just as with escape-avoidance conditioning, we now have a method for investigating the effect of other variables. One early use to which this technique was put was the analysis of the role of the various senses in the way the rat learned to traverse the maze. The influences of sight, sound, and finally smell were eliminated, and it was found that rats deprived in this way showed very little ability to learn. Smell was especially important in learning for animals already deprived of vision.

Variations in the basic technique have been legion. The various degrees of complexity of the maze itself in terms of its patterning of true and false paths – the blind alleys – have been touched on already. Then the influence of various guides to assist the animal in learning the route has been studied by indicating it by various coloured doors or lights of varying brightness. This has made possible a study of the capacity of the rat to recognize and utilize such clues in maze learning. This in turn has made possible a more refined study of its sensory capacities in terms of what different colours and degrees of brightness could be used with effect. Much of this work has been done using the basic enclosed floor maze with a covered top, but it is possible to elevate the whole pathway so that the rat has to find its way along the top of what is essentially a series of wooden walls, and this is especially useful in cases where it is necessary to study balance and co-ordination of movement, for example after the administration of drugs. Then again it is possible to utilize the basic maze-running approach but to vary the lure or incentive used. Thus the female rat may be separated from her litter, which is confined in the goal box for her to find when she has run through the maze; or a female in heat may be confined there instead and used as an incentive for male animals to learn the pathway through in order to reap the reward of sexual satisfaction. Even simply allowing an animal to explore the maze with no apparent reward at all has been shown to be sufficient to generate some learning under certain circumstances. Instead of attracting the subjects to the goal by positive lures in this way, it is possible to reverse the whole procedure and make the goal the means of escaping something, rather as one side of the shuttle box discussed earlier provides an escape from the shock on the other side. With simple patterns such as a T-shape, the floor can be made into a grid and electric shock used to motivate escape. Alternatively the whole maze may be put into water so that rats – who dislike getting wet – now have to wade or, if it is deep enough, swim through water to reach a raised platform at the end. The rat may even be required to swim under water through the maze by roofing the whole thing in below the surface of the water and only allowing it to come up to breathe at an exit in the goal box.

Two other methods of studying long-term behaviour motivated in some such way as the food-deprivation technique must be mentioned. They are the Skinner box for rats and mice and the Wisconsin General Test Apparatus for monkeys. The Skinner box is named after an eminent American psychologist and has been very widely used. Essentially it is a simple bare metal box in which the hungry animal is detained, and the novelty lies in the way the animal learns to work for a food reward. A little lever projects from the wall, and this is hooked up to a mechanism outside the box in such a way that every time the animal presses the lever, a tiny pellet of food is delivered to a trough inside the box. This is not a difficult feat for the rat to learn, and the advantage of this arrangement lies in the numerous variations which can now be introduced. The food reinforcement can be made intermittent, so that the rat has to press the bar, say, twenty times instead of once to get a bite of food, or it may receive this reinforcement at a fixed interval of time, say every minute. The extinction or waning of the habit of pressing the lever when no more food is forthcoming can also be observed, and it was soon found that the habit is very strong, especially after intermittent reinforcement of the sort mentioned above. Now, as in maze learning, many variations of this basic technique are available. Water may be used as the reward for lever pressing by the thirsty rat, or a blast of warm air for a cold one, or a period of light for one kept in the dark – or even a period of dark for one kept in the light. A recent innovation has been the delivery, direct to certain areas of the brain by implanted electrodes, of a tiny electric shock after each press on the lever. Is has been found that electrodes implanted in certain areas give so strong a reward that the rat will go on pressing the lever indefinitely in order to stimulate its own brain in this way, which presumably gives rise to pleasurable sensations. Now all of these various techniques are themselves of great value in other ways and have been especially useful in recent years in the determination of the effects of various drugs thought to be useful in the treatment of human mental disorder, as discussed in Chapter 7. One rather complex way of using the lever-pressing technique is to see how it is affected by other things – irrespective of the way you go about getting the rat to learn it. That is to say, once learned, here is a piece of behaviour which is stable, and convenient for study. The recording of the lever presses and the delivery of the food can all be made automatic, so that permanent records of many animals can be made simultaneously without constant experimental attention, something rather difficult to arrange with mazes and other learning devices. Thus, a simple bar-pressing habit based on a water-drinking reward may enable us to study a fear reaction. What is done is this: A conditioned fear or 'anxiety', as we may call it, is set up to a signal such as a buzzer in the same way as in escape-avoidance conditioning, but without allowing the rat to make any escape response. The result is that it will be afraid of the buzzer and show the usual ratty fear symptoms; that is, crouching, urinating, and defecating. But if we want to get a finer measure of this anxiety, we can use lever pressing and see what happens when we sound the buzzer while the rat is actually at work pressing for its water reward. The extent to which the lever pressing is interfered with (other things, such as the strength of the lever pressing habit, being equal) will indicate to us the intensity of the fear of the buzzer. Thus we can exactly measure the anxiety in terms of the reduction in the number of lever presses below the normal.

The Skinner box has been adapted for birds, and has proved especially suitable. Instead of pressing a lever the bird pecks at a target on the wall facing it. This is mounted in such a way that the force of the peck operates an electrical connexion, which, as with the rat, activates a hopper mechanism to deliver a food grain. Birds, especially pigeons, have been extensively used in this way, and techniques are available for doing complex experiments which formerly it was not thought possible to do with animals lower than man. Thus, the sensitivity of the bird's eye to light can be measured very accurately in the following way. The bird is first trained to peck at a target which is translucent and illuminated from the back – the bird in the Skinner box being in the dark. The bird is then trained not to peck at it when the light is turned off, and it learns to avoid doing this because pecks under these circumstances bring no food reward. Then a device known as an 'optical wedge' is introduced into the system, so the light is dimmed a little automatically each time the bird pecks. Now there comes a time when the light is so dim that the bird can no longer see it, and consequently it stops pecking, and by doing so it tells us in a behavioural way that it can no longer see the light. After a certain time the brighter light is restored and the process starts over again. Now, as is well known, a complex chemical reaction goes on in the eye which permits us to see better in the dark the longer we have been in it, and the same is true of birds' eyes. So we can trace the change in the sensitivity of the pigeon's eye in this way, because with each try more and more of the wedge is required to dim the light to make it stop pecking. In this way a curve of the increase in sensitivity can be plotted for this bird which is quite as accurate as that for humans, and just as certain as if it were possible for the bird to speak to the experimenter and say 'Now I can see it' and 'Now I can't'.

The Skinner box can also be used for larger animals such as monkeys, but most of the work on these species has been done by means of an ingenious apparatus known as the Wisconsin General Test Apparatus. Here again there is no especial novelty in the arrangements made for caging the animal: its own home cage can be used and the apparatus moved up to the bars of it, and the animal can do what it has to by reaching through them. Basically the device is a horizontal shelf which has two or more shallow indentations in it. These are for concealing the food reward – a few raisins or half a peanut will suffice for monkeys – and on top of each little trough is placed an object which conceals the well completely, so that the animal cannot tell whether or not there is anything hidden underneath it without displacing the object. The experimenter baits the desired object behind a screen which he can let down in front of the animal's cage so that the monkey cannot observe this phase of the process. Then it is lifted up and the monkey has to make the correct choice to get the food reward by pushing aside one of the objects from on top of the well. The problem might be to discover if the subject can distinguish round shaped objects from squares. Then all sorts of different sizes and colours of circles and squares would be used, with the correct round shape having the raisins underneath sometimes on the left and sometimes on the right of the square, which always covers an empty well. The animal readily learns the procedure of reaching out to get food, and a normal animal would very quickly learn to solve the problem posed above. But, as in all good tests of animal learning, there are many possibilities for variations in the basic pattern which can be used to advantage for special purposes. For example, there can be three food troughs, and the so-called oddity test can be applied. In this the animal always has to choose the one object which is different, in preference to either of the two which are the same. Thus on one trial an animal which has learned to do this might choose a round shape in preference to either of two squares, but on the next trial he would be correct if he were to choose a square in preference to either of two rounds. Or, again, there might be two green squares and one red – in such a case the red one would be the odd man out and so would be the correct choice. The effect of various experimental treatments (drugs, brain surgery, etc.) can be discovered by studying their effect on tasks of this kind, which can be made as easy or as complex for the monkey as is required by the experiment. Thus, it may be that the effect of brain surgery is expected to be quite severe, so there would be no point in seeing if the animal could retain the power to learn a very complex differentiation of solid shapes afterwards – it might be that all it could manage to do would be to learn a very simple task like the square–round differentiation mentioned earlier. On the other hand, the effect of some drug or other to be tested might be very subtle, and would only be observed when we require the animal to do something very difficult for it, and demanding a lot of concentrated effort on its part. Easy tasks would not be affected at all.

Finally, and in contrast with some of the more elaborate methods discussed in this chapter, much may be learned from simple techniques involving the manipulation of objects which are analogous to those tasks used in testing the intelligence of children. One such task requires the two-year-old infant to build a tower of blocks. Apes and monkeys can do this if a lure is suspended out of reach above the floor but directly in sight. With suitable material available for forming a tower a chimpanzee will readily solve the problem of reaching the food. However, there is as yet no recognized series of graded tasks comparable to those employed for humans and used for testing the intelligence of apes or other animals in the laboratory.

4

The Inborn Behaviour of Animals

IN this and the following chapter we shall consider some of the
findings which have resulted from the use of the field and
laboratory techniques discussed in previous chapters. These
findings fall into two broad categories – those relating to inborn
or innate behaviour, and those relating to acquired or learned
behaviour. This chapter will be primarily concerned with the
former, the innate or inborn; however, it is important to
realize that this is to some extent an artificial distinction,
adopted for convenience, and probably does not reflect accurately the real state of the world in which animal behaviour
is of such importance. As is often the case in science, the effort
to talk about it in terms other than the technical jargon of the
scientist himself frequently results in some degree of distortion.
And though it has been possible to avoid this difficulty throughout most of this book so far, it must be faced here in an especially
difficult form. The point is that there are few aspects or items
of behaviour which are solely innate, or solely learned. Most
are due to both inborn, genetic endowment – heredity in a
word – and to acquired, developmental forces – environment,
in another word. For example, the song of most birds is a
characteristic of the species which differs but slightly as between individuals. But does this form of communication have
to be learned afresh in each generation, or does it emerge fully
formed? The answer is neither: in song birds the elements of
the song are innate, but the actual form it takes is learned. This
has been shown by careful experiments in which young birds
have been hand reared without any possibility of hearing their

parents or other birds singing, and yet have produced a recognizable though primitive form of the appropriate song pattern for that species. Thus a relatively large element of learning may be said to enter here. Another example which has been much studied is sexual behaviour in lower animals. Do rats, for instance, instinctively know how to copulate when mating, or does this have to be learned – if not from other animals, as in the case of bird song, at least by a process of trial and error in which improvement is due to practice? Here again, the answer is that elements of both environmental learning and inborn hereditary endowment are involved. The male rat, reared in complete isolation, will mount a receptive female and make the appropriate pelvic thrusts to introduce its penis into the vagina, but the performance will be clumsy and less well integrated, each movement with the next, than it will be after practice. In this case, therefore, the contribution of learning may be said to be small. And so we could go on. Yet, nevertheless, there is a justification for the separation we are implying in the distinction between the subject matter of this chapter and the next: as we have seen, certain kinds of action can be regarded as being more closely determined by heredity than others, which in turn can be more clearly seen as environmental. Moreover, there has been a certain separation between the two in the way they have been studied and in the people studying them. The study of instinctive, largely unlearned, innate behaviour has been more often than not the concern of scientists who are primarily zoologists and who are often distinguished by the name of 'ethologists', whereas the study of individually acquired or learned behaviour among animals has been the special province of psychologists, who do not have a special name for themselves, unless it be animal psychologists.

Furthermore, it is difficult to generalize about the role of innate actions in animals – species vary greatly. It is important to understand clearly that when we use abstract nouns to describe behaviour – like instinct, in this case – what we are doing is to sum up the present state of knowledge regarding the facts about the way certain animals have behaved under certain circumstances. It is clear therefore that the title of this chapter is somewhat over-ambitious – it would in fact hardly be possible to summarize within the compass of this book all that is known about the inborn behaviour of all the species ever studied, so what I shall do in this chapter is to illustrate the way it is evidenced in certain species only and at certain times.

In general, the importance of innate behavioural organization recedes with advancing age. Obviously, it is biologically more important that newborn and young should be protected by instinctual behavioural responses to dangerous situations than older animals, who will have had time to learn appropriate avoidance actions. Therefore, in the examples which follow of what we know about innate behaviour, it will be seen that the innate, inborn, somewhat rigid, instinctual elements, where these have been adequately established as such, decrease with increasing age.

The start of life, for many species such as birds and reptiles, occurs inside an egg with solid walls, and their first problem is how to get out. Evolution has equipped most baby birds with a rather large and rigid beak which is well developed for the purpose but which would be useless without the propensity to use it properly, that is, to behave in the biologically appropriate way. This is a striking example of the importance of the interrelation between structure and function, and shows that the behaviour has evolved no less than the physical structure which it employs. This beak is brought into play only after the chick has started breathing through its lungs, and the watery contents of the egg, and consequently the egg shell itself, have begun to dry up. After the initial cracking or 'pipping' of the shell, the chick's neck begins to stretch and its legs to push against the shell, so that it splits the shell into two parts and emerges.

This behaviour is clearly innate; there is no possibility for imitative learning to have occurred from anyone else within the egg. Practice may have improved it a little, in the sense that the first attempts at chipping a way out may have been less successful than subsequent ones, so there is room for some trial and error learning. This, however, is made less likely by the mechanical nature of the act of breaking the shell – the first puncture must be the most difficult, thereafter the task becomes easier.

Some mammals may almost be said to go through two births. These are the marsupials, which lack a placenta. The Australian kangaroo is the best-known example. The young are born, and then remain for a further period in a pouch on the mother's abdomen, which acts as a sort of external womb and protects the young for a further period of development, after which the young emerge for the second time. The young are born in an extremely undeveloped and helpless state. Despite this lack of development, they have to undertake a highly important piece of behaviour at once in getting to the mother's pouch, where there is protection and nourishment in the form of milk from nipples which are located within it and on to which they latch very firmly.

The American opossum is a marsupial which has been especially studied in this connexion. The young are born only $12\frac{3}{4}$ days after fertilization and emerge very unformed. Nevertheless their forelimbs are proportionately large and tipped with claws. Thus equipped, these blind and otherwise practically helpless young find their way to the pouch, where they spend another eighty to ninety days before they are finally weaned. The mother licks the vaginal area as the young are born, and the babies crawl their way up her wet fur to the pouch. There seems little doubt that this

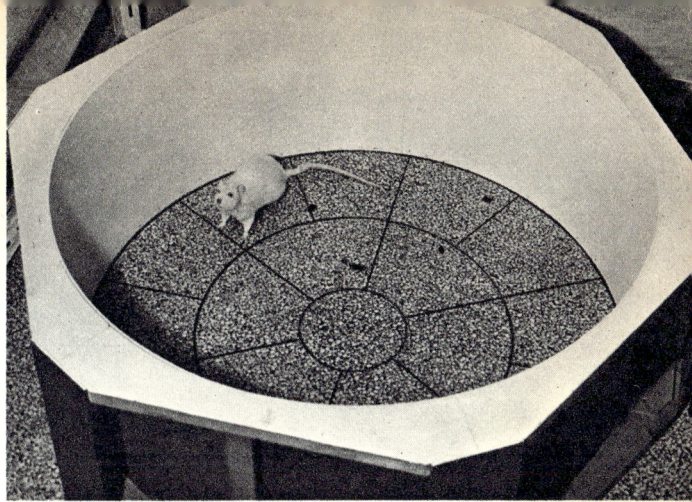

1a. The photograph shows a rat in the open-field test of emotionality (see pp. 51–2). The animal is taken from its home cage and exposed in the arena shown to mildly frightening noise and light stimulation. It responds by eliminating faecal pellets, several of which may be seen in the photograph, and by exploring the arena. The lines marked on the floor allow a measure to be kept of this activity.

1b. A rat making an avoidance response inside the escape-avoidance conditioning apparatus (see pp. 57–9). The buzzer has just sounded and the animal is running through the door in the partition separating the two grids. The rat learns from experience that the buzzer heralds a shock to the feet if it stays in the same compartment; so it runs into the other. Sensitive relays respond to its weight and turn off the buzzer, preventing the shock from being delivered.

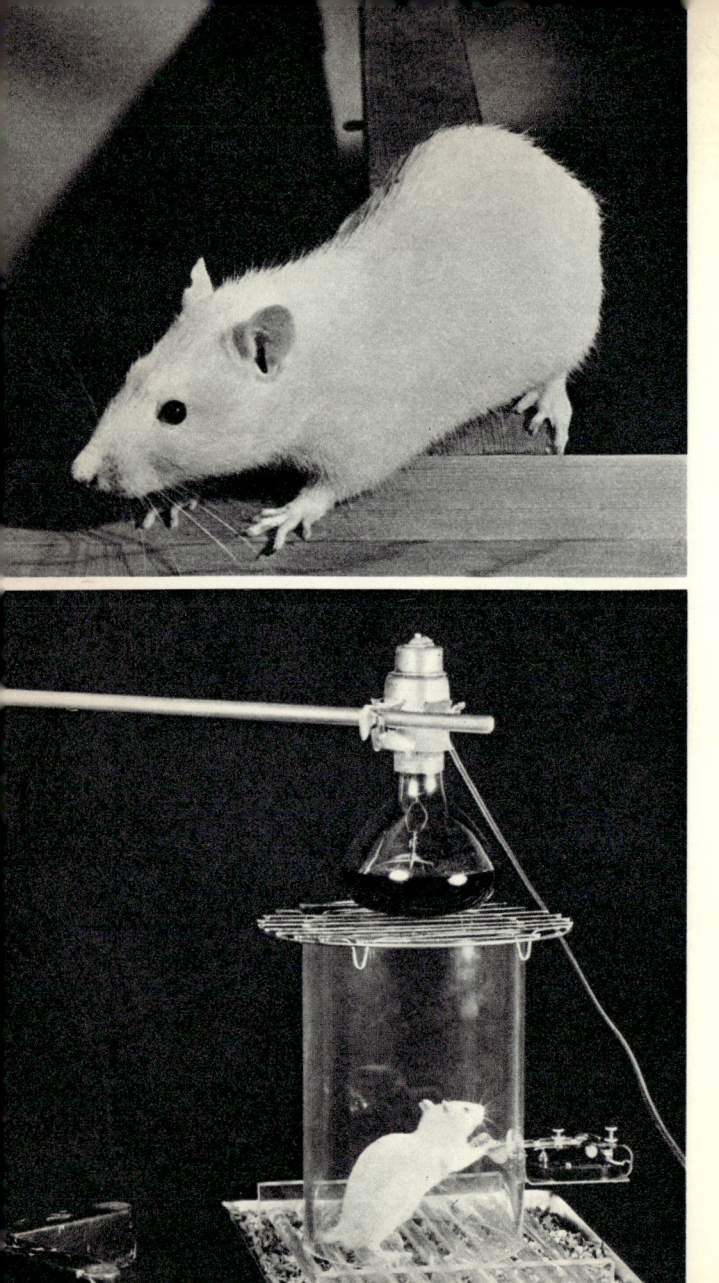

3. A rhesus monkey working at an oddity problem in the Wisconsin General Test Apparatus (see p. 67). (*Above*) The monkey has just pushed aside the round-shaped object to expose the food well it was covering and is picking up its food reward. The two square-shaped ones are ignored, as it has learned always to choose the odd man out.

(*Below*) Here the same monkey is choosing the square-shaped object in preference to the round ones, thus demonstrating its grasp of the principle behind the oddity problem (see p. 67).

2a (*top left*). This rat is making a turn on an elevated T-maze (see p. 63). Food is located at the end of the wooden wall forming the arm to the left of the picture and none at the end of the arm to the right. The animal learns to make the appropriate turn to receive its food reward.

2b (*bottom left*). The photograph illustrates an arrangement for learning based on heat reinforcement (see p. 64). The apparatus shown is located inside a cold chamber and the rat is confined in the transparent cylinder within which is a lever. This is connected to a switch outside the cylinder which controls the infra-red heat lamp fixed above. Depressing the lever turns on the lamp for a few seconds, and the rat readily learns to work for warmth in this way.

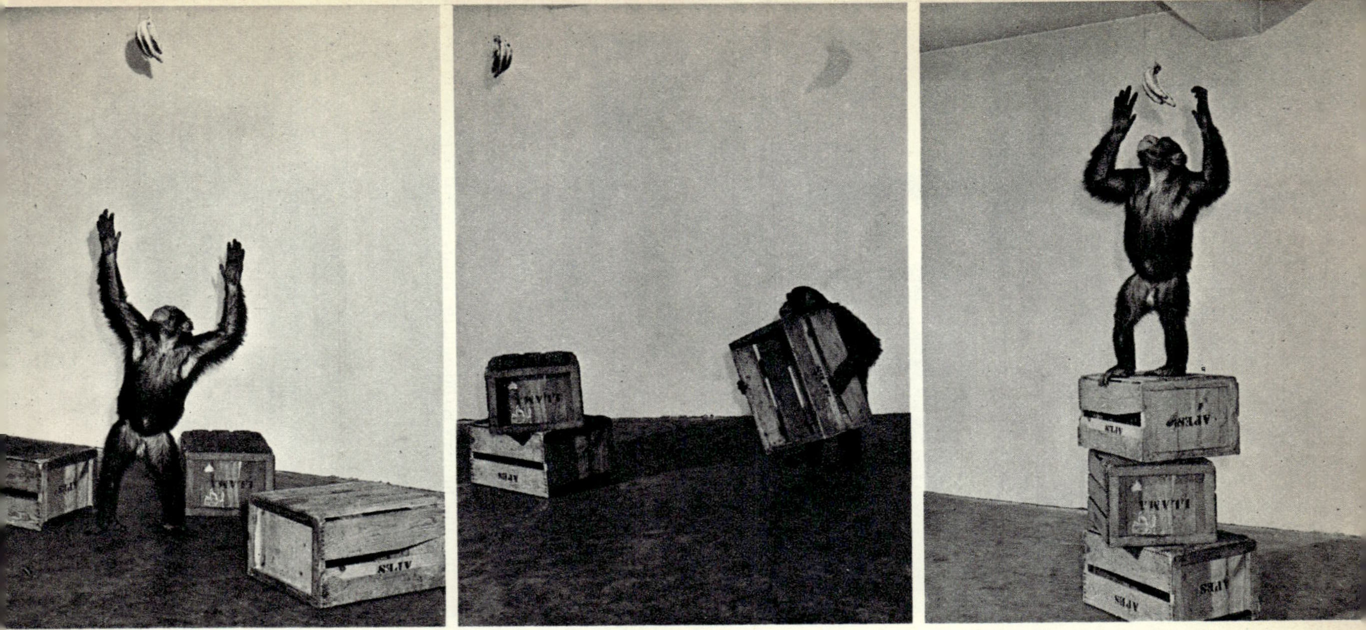

4. (*Left*) The chimpanzee has seen the lure and fails to reach it from ground level (see p. 68). (*Centre*) He progresses towards a solution to the problem by building a tower of two boxes, and goes to fetch a third. (*Right*) The tower of three boxes is completed and the chimpanzee is about to claim his reward.

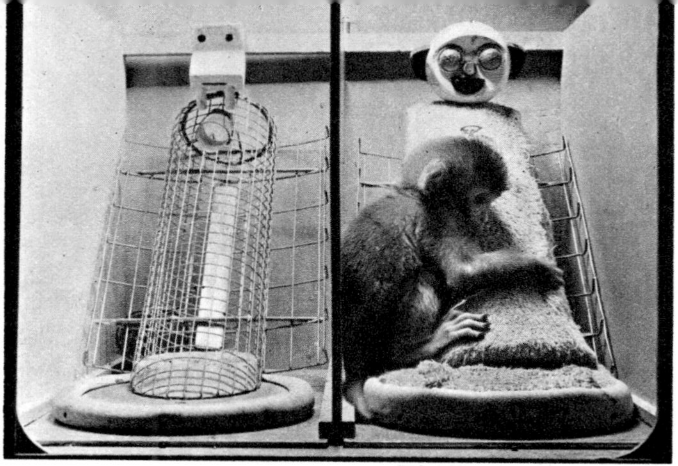

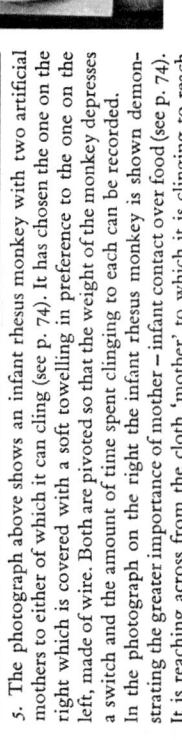

5. The photograph above shows an infant rhesus monkey with two artificial mothers to either of which it can cling (see p. 74). It has chosen the one on the right which is covered with a soft towelling in preference to the one on the left, made of wire. Both are pivoted so that the weight of the monkey depresses a switch and the amount of time spent clinging to each can be recorded.

In the photograph on the right the infant rhesus monkey is shown demonstrating the greater importance of mother – infant contact over food (see p. 74). It is reaching across from the cloth 'mother' to which it is clinging to reach the milk supplied, in this case, by the wire one.

6. These seven goslings have been imprinted on the man and follow him as if he were their mother (see p. 73).

7. Looking down into an underwater discrimination apparatus (see p. 98). The photograph shows a rat about to emerge from the water after having made a correct turn to the brightly lit side of a Y-shaped apparatus. The choice-point is at the top of the picture and the channel on the left is dark and the escape door locked. Rats readily learn to swim underwater and to make the correct brightness discrimination.

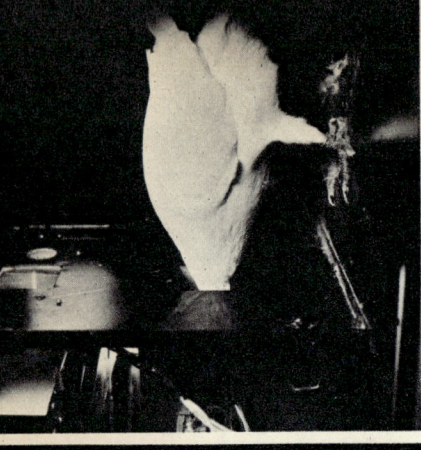

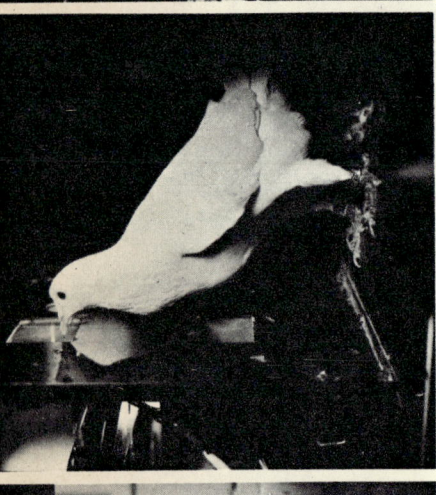

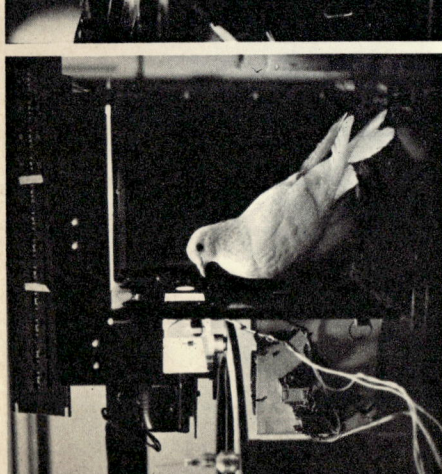

8. Pigeon performing an inspection task (see p. 133). (*Left*) A general view of the experimental arrangement for presenting the components passing one after the other through a brightly lit compartment on the side of the Skinner box in which the bird is confined. The square window through which the bird is looking forms one key for the bird to peck at when the components are defective (see next picture) and the round one next to it another for responding when it judges components to be good.

(*Centre*) The pigeon is pecking at the square key to indicate a defective component. If this is one of the five per cent of known

'duds' sent through by the experimenter, such a correct response to it will result in a food reward, but if the pigeon makes a mistake and pronounces it as good by pecking on the round window the lights are turned out for a short while. This constitutes punishment since it temporarily prevents the bird from responding and so earning its food.

(*Right*) The pigeon is obtaining its food reward for the correct response shown in the previous photograph. The food is delivered into a food hopper located below the keys.

behaviour, seen in the newborn at an especially early stage, is completely innate. But how it is accomplished is not completely understood. The young show a tendency to move upwards, against gravity, as do the young of some rodents such as rats, and this doubtless assists the process of finding a nipple.

Having been born, then, what is the next most fundamental thing for the young animal? One immediately thinks of recognition of danger and of security. With many animals security often lies in the person of one or both parents, and most often the mother. The mother is often either herself the source of food supply for some while, as in lactating mammals, or she leads to, or provides, suitable nourishment for the dependent young. Recognition of the mother is therefore of extreme importance, and in recent years there has been a notable advance in the understanding of how this comes about in many species, especially in birds. It has been shown that there occurs at a certain age a fixation of the young on the image of the mother – it has been given the name 'imprinting'. Once 'imprinted' – and the internal mechanism by which this is done is not completely known – the young will follow the mother and no other bird. This behaviour obviously has great survival value, and is of interest also in the perversion of it which can occur. As in many branches of the life sciences, pathology, or the study of things going wrong, can be particularly illuminating. It seems that there arises a specific readiness to be imprinted in certain species of birds at certain ages, and if the unusual – at least in nature – event occurs that the appropriate parent is not around at this rather crucial time, the imprinting will nevertheless occur, but on to the wrong object. Thus the spectacle may arise of, say, a brood of geese following a human and behaving in many respects as if he or she were their mother.

The infant–mother reaction has been studied experimentally in the laboratory setting by depriving newborn rhesus monkeys of their real mothers and providing them with substitute mothers made of various materials. These infants were bottle-fed, of course, and the baby monkey had to climb up on to the substitute mother in order to feed. Now it was found that, when allowed to choose between two kinds of substitute mother, the infants spent more time on models covered with a soft terry towelling than on ones in which the wire frame was exposed to the touch. This was rather as might be expected, the towelling being more like the natural mother's fur to which the infant monkey normally clings than the hard wire frame-work. But what was also found, somewhat unexpectedly, was that they spent more time clinging to the cloth even when they had to go to the wire mother for milk. That is, the com-forting effect of the cloth was more important in the artificial mothering process than was the fact of feeding – they preferred a soft cloth model even when it did not provide milk to a hard wire one which did! This innate preference can be thought of as being a protective mechanism in the sense that, though it is clearly important that the infant should receive its food, it can be thought of as less important for survival than clinging close to the mother in the early arboreal existence of monkeys and the like. Contrary to what is popularly supposed, tree-living monkeys, though undoubtedly agile, are not incapable of falling out of trees and injuring themselves. The study of wild, freely-roaming monkey troupes shows that a surprisingly large number of them have suffered bone damage (fractures and the like) of some sort or other – presumably usually through falls. Thus the ability and necessity for clinging to the mother's fur must be very strong in the infant monkey if it is to survive this, as well as other, hazards of its young life.

Mothers are important, it is generally agreed, and we have seen in the experimental work described above just how their importance can be analysed by changing some of the con-ditions which occur in the maternal relationship with their offspring. But just how important are mothers? Can they be dispensed with altogether? The answer would seem to be 'No'. The monkeys which were used for the experimental work referred to were followed into their adult life, and it very soon became apparent that one most important aspect of what is usually regarded as innate behaviour – responsiveness to the other sex – had been seriously interfered with. They showed no interest in mating, and no ability for it. Even paired with an experienced partner – that is, a male or female who had been known to mate successfully in the past – they still displayed their abnormality. This was so severe as to preclude many of these motherless monkeys from ever bearing progeny, while those that did mate more or less successfully and bear young turned out to be extremely poor mothers themselves, ignoring or even ill-treating their offspring. These examples indicate the profound extent to which an innate behaviour pattern – sexual responsiveness – can be modified by experience. Of course, the experience of being deprived of a mother is unlikely to occur among wild monkeys, for the infants would be unlikely to survive, though clearly natural selection must operate strongly against the inheritance of even slight tendencies towards these forms of abnormal behaviour in the wild. Animals which mated ineffectively or reared their young imperfectly would necessarily ensure that their own genetic make-up is represented in later generations less well than that of their more normal contemporaries. And so what must be inherited in animals like monkeys is not the mating and maternal behaviours themselves but the propensity to develop them, given the appropriate environmental experience at the appropriate time in the individual's development – that of being mothered, in this case. Thus the normal pattern of mating is a result of the interaction of the hereditary make-up of the animal and its own individual life history.

However, it is clear that, especially in animals lower in the evolutionary scale than monkeys, the whole complex of activities associated with mating and rearing the young has many important innate components which do not depend on individual experience for their emergence. We have seen in the descriptions of field observations of large groups of animals which live together in social groups how important some of these activities can be. For example, the territory, the area of space around the animal's home or nest which it regards as its own and which it will defend vigorously, forms an important role in pair formation, especially in birds, but also in other animals. Thus the remnants of territorial behaviour are seen in the dog, despite many centuries of domestication from its wild ancestors. This is the familiar canine habit of marking posts, etc., in its area with its own urine, often to cover up that of a stranger. Wild bears do the same thing with their excrement. But birds display this tendency in its most developed form. Probably many of us are familiar with the experience of a small bird battering at a window pane, often day after day, especially in the spring. It is not, however, trying to get in, though it may seem so. Investigation will usually show that there is a tree outside the window, and that the window reflects the light in such a way that the bird sees its own image when sitting on a branch near by. It then sees its own reflection as a rival bird on its own territory and attacks it, but of course, fails to drive it away as would normally happen, since intruders usually defer to the established owner of the territory and are eager to escape. And so the bird will resume the attack, time after time, day after day.

Courtship behaviour comes next after the establishment of a territory, and the song of birds serves in many species as an indicator of the occupation by a male bird of a suitable nesting territory into which it is trying to attract a female mate. Also surprisingly complex and stylized rituals have often been evolved in this connexion: the display of the peacock with its veritable fan of iridescent tail plumage is probably the best-known example of this, but there are numerous others which have been carefully documented by students of animal behaviour. Birds are especially rich in this kind of activity, and often, like the peacock, have special feathers and colourings which have been developed for the purpose. The same is less true of other animals in general, which are, unlike birds and insects, mostly colour blind.

The innate aspects of sexual behaviour have been much studied. The pathology of sexual behaviour can be observed as we saw in connexion with the monkeys mothered by machine, but normal sexual behaviour has been the more usual object of interest, or at any rate normal sexual behaviour as modified by the conditions imposed by the experimenter. This line of inquiry seems to have brought forth especial ingenuity on the part of experimenters. For example, the normal heterosexual behaviour of the laboratory rat has been studied in great detail, one technique being to wire the members of the pair up in such a way that sexual contact closed an electrical circuit, thus allowing an imperceptibly small – to the rat – current to flow which enabled the duration of copulation to be recorded electronically. And the question of the development of homosexual relations has not been neglected. It was studied by forcing male rats into making sexual approaches to another male – mounting and pelvic thrusts, etc. – by keeping them on an electric grid and shocking them through the feet, then turning the shock off whenever they commenced the homosexual behaviour. That is, a method of escaping the shock which the rats could readily learn was provided. In this way it was shown that rats could be taught to be homosexually oriented, and to reject females altogether. The reader may be wondering how the experimenter managed to avoid shocking the experimental rat who served as the subject of the advances. He was provided with little rubber boots to guard him against the shock!

Normal sexual behaviour in guinea-pigs has similarly been extensively analysed. Such behaviour creates especial problems of measurement, but these can be overcome by considering one sex at a time and using the other sex as a standard, and trying to make the conditions of stimulation as constant as possible. Thus, when studying males, a female known to be in heat was used, and when females were studied they were artificially brought into heat by an appropriate injection of endocrine hormones beforehand and tested for the arching of the back which indicates readiness to copulate. Then the behaviour of the male animal under test can be observed when brought into contact with a female by counting the number of movements in the various components of the sexual act in this species – sniffing, nuzzling, mounting the female, intromission, and finally ejaculation. But all this work was preliminary to the major question investigated, which was to find the way in which the various components of behaviour involved in mating are inherited. This was done by crossing two strains of guinea-pig which differ in their readiness to mate and then comparing the performance of the cross-bred offspring with that of their parents. In this way it became possible to show that, rather as expected, the genetic contribution was fairly important – in some cases well over half of the differences between the various groups could be attributed directly to hereditary agencies.

After sex come some consequences; and in nature mating sets in train a whole series of processes, both physiological, in the way the animal's body begins to prepare for the birth of the young, or the laying of the eggs, and behavioural, in the preparations the mother, and sometimes the father as well, make in getting ready to receive it. This is perhaps most conspicuous in birds where nests are made, and its hormonal (glandular) basis has perhaps been most thoroughly studied in doves. The female ring-dove's behaviour in mating, building a nest, and incubating the eggs is in turn controlled largely by a series of different hormones – one activity stimulating the appropriate hormone for the next and to some extent inhibiting or 'shutting off' the supply of the one currently circulating in the bird's bloodstream.

The way birds typically respond to all sorts of bits of appropriate material with which their nests may be constructed makes further analysis possible of the instinctive nature of such behaviour. The weaver-bird in South Africa uses a knotted strand of horsehair as a foundation on which to build a complicated nest of sticks. A pair was isolated and five generations of weaver-birds were reared by canaries, completely out of sight of any others of their fellow species and deprived of their usual nest-building materials. In the sixth generation, still in captivity, when they were provided with the right materials, they built typical weaver-bird nests – they even used the knot of horsehair. This striking example strongly argues for the completely innate basis of the birds' reaction, at least in this species, to the nest-building material.

Nesting behaviour can vary from one related species to another, and this allows a genetical analysis of its inheritance. There are, for example, certain species of lovebirds, a sort of small parrot, which carry nesting material, grass, leaves, etc., in their beaks to the site where they are building. Others carry it tucked under their wing. Now what happens when you cross these two kinds? Do the cross-bred ones use wing or beak? The answer is that at first they did neither: they picked up material in the beak, and then tried to put it under the wing, but halfheartedly, so that it had fallen out by the time they reached the nest. It took some three years for them to learn to carry it in the beak, which shows that they were able to profit from experience in that the inadequate behaviour, established genetically by the mis-mating of species practised by the experimenter, could in time be overcome.

Next in the normal cycle of behaviour comes maternal behaviour – the care of the parent for the young – which is yet another aspect of innate behaviour which has been thoroughly investigated in several species. We have already seen how deleterious the effects of its absence can be in monkeys, but in its normal manifestations it shows striking unlearned components. The devotion of the adult – both male and female – penguins to their young when they literally starve for many days before and after the young are born is a case in point. The tale of the meadow-pipit or other small bird whose nest harbours a parasitic cuckoo is familiar. Here the parasitic species has evolved in a way which capitalizes on the parental devotion shown by these birds in an opportunistic fashion, so that the spectacle may be seen of the two willing dupes of foster parents working hard to feed the monster in the nest now grown larger than both of them put together and which in the process has thrown their own young out of the nest! This may be, however, another case in which supernormal stimulation plays a part, just as it did with the case of the gull's eggs discussed in Chapter 2.

Parental solicitude among animals – or apparent solicitude, rather, for this seems an especially appropriate moment for reminding ourselves of the dangers of anthropomorphism and of seeing human motives and concerns in animal situations where they do not operate – can take many unusual forms. Among whales, and other species of mammals which have evolved into fish-like forms and returned to the sea, the young are born alive and depend upon their mothers' milk for early nourishment. But suckling an infant under water presents many problems – I doubt if even the most avid lady skin-diver in the current development of the sport has tried this. Not least of these problems is that the infant may drown if it sucks water into its lungs as well as the milk into its stomach. So the behaviour which has evolved is a neat ancillary to the physiological changes from the method of lactation in land mammals which have accompanied it. The milk is thicker, richer, and stronger, so the infant needs less of it, and the mammary glands have developed a pump-like action by which the milk is ejected. The mother whale's behaviour which assists this process is to surface and float on one side so that the nipples are out of the water for the young creature to suckle. All the physiological differences described make for a speedy feeding process, so the time taken for the whole process, when both mother and young would be on the surface and vulnerable to enemies, is shortened.

With the onset of maturity in the animal the importance of innate behaviour patterns wanes, with the exception of its sexual and parental functions. This is not surprising, for the propagation of the species is the fulfilment of its function. The role of experience in the survival of the individual thenceforward assumes a greater importance to the extent that it is difficult to detect any innate patterns of behaviour in older or senile animals. There are, however, many characteristics of the way the animal behaves throughout its life which are in part determined hereditarily, and are susceptible to the methods of analysis provided by the science of genetics. One of these we have already encountered in the hybridization and cross-breeding of birds and guinea pigs mentioned above. The other which has been used – though not very frequently, for it involves much effort – is that of selective breeding. This resembles the techniques employed by animal breeders in their efforts to improve a domesticated strain, often by seeking to make it approximate to some ideal type. However, in the laboratory, where this work almost inevitably has to be done, because of the need to standardize the environment as much as possible in order to limit non-genetic causes of variation, the usual procedure is as follows. A whole population is measured for some characteristic such as performance in a maze, and then those animals which have scores above or below a certain score are selected for further breeding as parents of the next generation. Sometimes both these things are done, in which case two new strains are developed from the original population; one strain which shows a high score on the maze, the 'bright' rats, and one a low score, the 'dull' ones. This procedure was used successfully with rats in a deservedly famous experiment in which two strains which differed in 'intelligence' were created during the 1930s. Their descendants are still alive today and are widely used for various experiments in which animals of varying intelligence are needed. Other experimenters using selection for different types of behaviour have studied the cage activity of rats and their emotional responsiveness to a novel situation, as described in Chapter 3. The interest of these experiments lies in the fact that, in general, if one can breed selectively for something, this is strong evidence in itself in favour of there being an important hereditary component in the way it is determined in the species. And by modern techniques of analysis it is often possible to discover how strong this component is, in relation to the environment, and how many units of heredity are involved. As we have seen, it is rarely the case that it is possible to be at all precise regarding the interaction of the innate and the environmental.

But still there are many characteristics possessed by the adult animal that are largely innately determined and which both affect its behaviour greatly and influence the way it senses the world around it. Indeed, one of the most fertile sources of human misunderstanding of animal behaviour, its motives and its operation, is our failure to appreciate that animals see, hear, smell, and feel things differently from the way we do. In addition, of course, many animals possess different kinds of sense organs: insects have delicate organs of balance, and fish possess the power to detect subtle changes in the water pressure on their sides. But for most animals the sense organs broadly resemble our own. The differences are principally due to the different structure of the sense organs, the different way they operate, and the differences in the relative importance of the information they transmit to the animal. Thus the predominance of vision in man and the other primates (apes and monkeys) is not general among many animals. Rats, for example, are not seriously hampered in learning the pattern of a maze if they are blind from the start. Dogs, as is well known, rely largely upon scent for their exploration of the world around them. But despite this lack of emphasis on the importance of vision, one of the most fascinating differences for the student of animal behaviour is the response to colour. Colour blindness is not unknown in humans, indeed approximately one in twenty men suffers some degree of deficiency of his colour vision, though the incidence is lower in women, this being a sex-linked genetic trait. But its existence has only been relatively recently recognized. (It was for a long time known as 'Daltonism' after its discoverer, John Dalton, the English chemist.)

It is commonly said that a red rag incites a bull to anger. It is almost equally often said that this is nonsense, because cattle are colour blind. Since I know of no study which adequately establishes the truth of either statement, and since this is a book which relies upon scientific evidence and not folk-lore, there the matter must rest! Rather more, however, is known about colour blindness in other species. Thus it is possible to show, by laboratory methods described earlier in this book, whether or not an animal is actually making a response in an experimental situation which depends upon its perceiving a colour or not. The rat has been shown not to be entirely colour blind, a greater sensitivity to red lights being especially detected. The method used was to train rats to jump from a platform into one of two boxes mounted side by side in front of it. The 'correct' one contained food, the 'incorrect' one had a false floor, which gave way under the rat's weight, allowing it to fall into a net below. The animal thus had to learn to jump always to the side displaying the 'correct' design, irrespective of whether it was presented by the experimenter on the left or the right.

Thus one rat was trained to jump to blue and avoid red, another to jump to red and avoid blue. The lights shone through coloured filters at the backs of the boxes, and the filters could be switched from side to side. Elaborate arrangements were made to prevent the rat from learning a difference between the lights because of differences in their relative brightness, as opposed to colour, which can happen. A brightness discrimination, as we shall see, is relatively easy for the rat to learn, and without excluding this possibility it would be erroneous to conclude that the rat was learning which side to jump on the basis of an innate colour sense.

These then are some examples of innate endowment in animals which can be detected by using behavioural responses in a laboratory setting. But these are somewhat artificial and indeed are selected solely for use as a suitable indicator of the animal's sensations at the moment. Are there no other tendencies which are of importance in determining the mature animal's behaviour? There are, though most of them derive in some measure from the sexual and parental functions described earlier. Thus many animals show a social tendency, that is, they spend the major part of their lives in groups or herds. In these groups a dominance order may be observed, so that the more dominant animal may take precedence over the others in choice of food or of mate or in leading the group to a new place. This is often the role of the largest and strongest male, and the leadership is established and maintained by combat. But this is not always the case: in domestic sheep, for example, the leader is the oldest female, which may seem a strange choice, but which comes about like this. The lamb keeps near its mother's side and to her rear at first, for obvious reasons. As it grows it becomes more independent of her, but if at all disturbed or fearful, as when the flock moves – usually a time of some slight stress – it resumes this position. Now next year, the female lamb, now grown up, has its own lamb, which in turn trails its mother, and so on, so that the whole flock in time is congregated behind the most senior mother. When she dies or is removed, her daughter is automatically promoted.

Social behaviour in primate colonies has been intensively studied both in the wild and in zoos by the methods described in an earlier chapter. In several cases colonies on small islands have been observed: these have the advantage of ensuring a limited territorial range of particular groups, which are consequently easier to identify. Also no new admixtures to an island population occur without the intention or at least the notice of the research workers. One important series of studies was done on an island in the lake forming part of the Panama Canal; another on a small island off the coast of Japan. Here the Japanese monkey has been studied in great detail by the provisionization technique described in Chapter 2. The way the food provided is taken by the various individuals in the group or band of monkeys reveals much about their relative standing in the group – their social dominance as it is called. Wheat was placed on a box and the order in which the monkeys permitted each other to mount the box to get the food enabled a tabulation of the respective rank of the monkeys from the 'top dog' downwards. Such rankings are surprisingly stable, and at first the young monkeys have the status of their mother in the hierarchy. A single individual introduced experimentally was not integrated into the group and five years later was observed to be still leading a solitary existence.

More difficult are the intensive studies of gorillas in the wild. This is because the gorilla is to be found only in very limited areas of West and Central Africa in dense forest, and is extremely shy. On the other hand its study is of extreme importance, since of all the primates it is in some respects closest in resemblance to man. The gorilla is one of the few primates that makes a nest, though, of course, many animals other than birds do so. Thus the domestic rat and the mouse will both construct nests, given the availability of suitable material. (This is an innate tendency which manifests itself in the presence of such materials and which has been much studied as the prototype of instinctive behaviour.) But the nests of the gorillas are primitive affairs compared with the rather elaborate productions of some species of bird, which we saw earlier. This is not altogether surprising because they are often used for one-night stands, that is to say, they are constructed on the ground or in the branches of trees for sleeping in overnight and are left the next morning when the gorilla family moves on.

Innate tendencies to use natural materials in nesting are also met with in the few examples of tool use in animals, other than insects, which show a great variety of reflex responses to in-animate objects which may be used in constructional purposes. One species of insect may be said to use physical material as a tool. This is the burrowing wasp Ammophila, which employs a small pebble to pound down the soil over the burrow where it has laid its eggs. This obliterates any traces of disturbance made by the digging and so completes the concealment, and hence protection, of its nest. A species of bird is known to use a tool. This is Darwin's woodpecker finch, one of several species of finches which are to be found in the Galápagos Islands off the Pacific coast of South America. In the absence of competition from other species of birds, the finches have evolved into many different forms, to fill the various environmental opportunities the islands provide. One behaves like a wood-pecker, but, lacking a beak adapted to gouging out insects from the bark of trees, it uses a cactus spine for the purpose, holding it in its beak and dropping it as soon as it wants to seize the insect thus disturbed. In areas where there are no cactuses, it selects a twig of suitable size and flexibility. Another bird, the greater spotted woodpecker, uses a cleft in a tree trunk as a vice in which to wedge pine cones and oak apples while it picks out seeds and insects.

One authenticated case of the use of natural objects in a tool-like way in nature by mammals has been reported. This is the feeding behaviour of the sea otter, which dives to the sea bottom to collect various shell-fish as food. One of these has an especially hard shell, and the otter therefore brings up with it a flat stone against which to crack it. It lies on its back on the surface of the water with the stone anvil on its chest and brings the shell down with its forefeet on to it with great force to release the fleshy contents, which it then scoops out with its front feet and conveys to its mouth – or to that of its pup if it is a female with young. The young spend much time riding on the mother in this way, for sea otters usually swim and even sleep on their backs in the water.

Mammals also show a few examples of the elaborate use of natural materials. Perhaps the most striking innate behavioural response is seen in the dams constructed by beavers. They begin with the apparently simple action of felling trees by gnawing the lower stem away. They then dam a stream, forming a pool of deep water. This has the effect of preventing the water from freezing during the winter, and the beavers' food, the bark of tree-branches which are stored at the bottom of the pool, is thus protected and, most important, is accessible to them in a way it would not be if the water were frozen solid. Near by, on a bank or shoal, the nest of bark and twigs is formed. Then this is surrounded by a wall of similar material, reinforced by mud and arched up over the top to form a dome. This is the beavers' lodge. When this damp structure freezes solid, it forms a warm house for the animals, not unlike the Eskimoes' igloo, and protects them against bears and other predators. The group in residence has not been known to exceed fourteen in number, and is usually less. This maximum of fourteen for the colony is set because the beaver has a litter of not more than six, and the young stay with the parents for their first year, so that these yearlings plus the new litter, plus the parents constitute the colony in the lodge during the winter. The food, stored in deep water as described, is brought up to the lodge through a plunger hole in the ice, which is located within the protecting dome. The wood remaining when the bark is eaten is used to repair the nest and the dome above, or to increase the height of the former if the water level rises. The colony inside the lodge generates sufficient heat to melt a ventilation hole at the top of the dome.

Now all this activity may have a profound effect upon the drainage of an area, as the dams may impound a lot of water. Indeed, in some parts of America, beavers have been deliberately introduced as a water conservation measure, with some degree of success. The complex series of actions necessary to complete the beavers' elaborate protective structure and to provision it have not been studied in any detail but are doubtless dependent upon hormonal, glandular changes in the way that the nesting and egg-laying of the dove is, as described earlier. But they are also subject to social stimulation and possibly learned experience, for individual animals which do not cooperate in the dam-making processes may become isolated and return to the more primitive way of living in burrows.

5

The Acquired Behaviour of Animals

ENOUGH has been said in the previous chapter to make it clear that the term 'acquired behaviour' is more of a classificatory device for ordering material than a statement of fact about the way in which behaviour is determined. The interaction of environmental and hereditary forces is especially complicated in the case of acquired behaviour because it so often seems among higher animals that the disposition to act in a certain way in response to certain environmental conditions is inherited, whereas the exact mode of action – the way in which it is done – is largely determined by the circumstances of the moment. Thus, if a rat has learned a certain pattern of maze to attain a food reward, it may be immaterial to the animal's success whether it runs through a dry version of the maze or swims through the same pattern immersed in a tank full of water. Now this example suggests, and rightly so, that what we shall be concerned with in this chapter is behaviour which is learned – that is, some way of acting which the animal did not display when first presented with the conditions in question, but later, and, perhaps after repeated experience with them, now shows quite regularly. Learning has been intensively studied by many students of behaviour, because of the crucial role it plays in humans. The human being is for ever learning, adapting his behaviour to changing and often novel sets of environmental conditions. Trivial examples are seen in the need to learn new faces and new names. More important is the acquisition of new skills, like learning to type, or the crafts of a new trade; and of course, for children, the whole educational system and the whole pattern of socialization, that is, the development in them of behaviour acceptable to the society in which they live, is based on learning. Yet calling this acquired behaviour must not lead us to think of it as essentially accidental and dependent upon chance concatenations of appropriate conditions: it may be that in the nature of things all members of a particular species may be exposed to the same set of conditions at about the same time in their development and so pass through the same learning process and consequently display the same sort of behaviour at a later stage in development. It is in this way, incidentally, that identity of behaviour can arise among various members of a species and is yet not innate, though at first sight may seem so. A further consideration which must be noticed is that while it is sometimes possible to see a definite physical basis for much behaviour we have been calling innate – such as a special sensory or response apparatus in the animal – it is often not possible to say the same about learned behaviour. Thus it is clear that a rat will, say, feed its young by allowing them to suckle rather than by regurgitation as many birds do, because of its possession of nipples which are obvious upon inspection. But we are not able to tell by inspection that it knows the difference between a buzzer and a bell, and yet there probably is some definite difference in terms of a change in the nervous system, especially the brain, as between two rats, of which one has learned the distinction and the other has not.

The physiological basis for learning is as yet not completely understood, and indeed it would still not be possible to distinguish between two such animals even if they were killed, dissected, and their bodies subjected to the most refined analysis at present available to science. But in a very real way they can be considered as different animals – the rat subjected to an experiment arranged in such a way that it was highly important for it to learn the difference between bell and buzzer, for example, to avoid an electric shock, might literally never be the same again afterwards. It might retain the capacity for reacting to the bell in a special and observable way which was quite different from that of an animal which had not been through the same experiment – what is called in laboratory jargon a 'naïve' subject – and which as a consequence might be indifferent to bell, buzzer, or whatnot. Of course, it might be objected that forgetting can occur in the trained animal, so that it can be restored with the passage of time to the status of a naïve animal. This is, of course, true, and many studies of animal memory have been based on just such assumptions. But it is also true that there are sensitive methods which can be used to show that even when reaction to a signal of the sort we have been considering is no longer apparent evidence can still be detected which shows that the effect of the original learning is still there in the animal. Thus, for example, it may be that an animal which had been trained in this way will be re-trained again faster than one without the prior experience, thereby revealing the effect of that experience, despite the passage of time. What exactly is learned – the subtle alteration in chemical structure of various nerve cells in the brain of the animal, the nature of which is at present unknown – can conveniently be thought of for the moment as a memory 'trace'. This in turn leads to the question, how is this 'trace' acquired? Here too the answer is as yet not known, but various theories have been advanced to account for the facts which are known about the learning process. But though no final answer can be given there is one set of considerations which many workers think affords a clue to the way organisms acquire learned behaviour. This lies in the process known as 'conditioning'. We have already met the term in connexion with escape-avoidance conditioning, when it was pointed out that it had its origin in the word conditional – referring to the process of making a natural innate reflex occur in response to an artificial signal, dependent – that is, 'conditional' – upon whether or not the two had been paired together a number of times in the animal's experience. The classic experimental work using this approach is that of the great Russian physiologist, Pavlov. He used the salivary reflex of the dog, an innate, natural response to the sight of food in this – as in other – species, and paired it with a laboratory signal, often the sound of a bell. The dog was restrained in the laboratory in a type of harness, and the saliva was measured by attaching to the animal's cheek a tube through which it was collected. Then the bell was rung, and the food presented by a mechanical device which caused a plate to swing out from behind a screen. At first, the bell had no effect; the dog only salivated when the food arrived. But in time the dog learned to anticipate the food in response to the bell signal, and so the salivation commenced before the food came. Further practice of this pattern caused a very stable situation in which the dog consistently responded to the bell over long periods without being rewarded by the presentation of food. Thus we can now say that the bell had become a conditioned signal for the salivary response, and the salivary response to the bell is spoken of as a conditioned response. This simple principle was elaborated by Pavlov and subsequent workers in many different ways. For example, if one sort of signal is always followed with the food, and another, such as a light flashing or even a different sort of bell, is never rewarded with food it becomes possible to show that the dog can discriminate between the two, in that he will regularly salivate to the first and never to the second. Thus one signal is thought of as the positive conditioned stimulus and the other as the negative. In this way a whole complex of learned responses can be elaborated in one animal, and many aspects of the way the responses are best learned and the way in which they are forgotten can be studied.

It was this kind of experimental work which led to the

formation of theories of learning, which sought to account for the much more complicated responses to the environment learned by, say, a rat running in a maze, or, and more especially, humans in their everyday lives. It cannot be said that the solution to the problems posed by the many facets of learned behaviour has as yet been brought within sight by this or indeed any other formulation, but many scientists have found it helpful to think of learning in terms of a series of conditioning procedures, occurring throughout the life of the individual animal.

Like innate behaviour, which was considered in the previous chapter, acquired behaviour in animals (of which conditioned responses are but one example) is too large a topic to review in this chapter. So, as before, we shall take representative findings relating to acquired behaviour from various kinds of animals at different stages of life in order to illustrate the sort of scientific work done in this connexion.

As before we shall start our survey by a consideration of the possibility of behaviour being modified before birth. Here we shall be concerned with behaviour which would not otherwise be observed in animals after birth had it not been for some experimental treatment being given to the mother during her pregnancy. It is possible to show that the mammalian foetus in general will respond to external stimulation, and some pains-taking work has been done on animal foetuses delivered by surgical procedures at different stages of gestation. But this has merely demonstrated the presence of certain reflex responses and not the possibility of modifying behaviour as such.

But if a female rat is put in an escape-avoidance conditioning box and taught to fear a signal, such as a buzzer or a light, by giving it a mild electric shock each time the signal occurs, and it is then mated, we can go on exposing it to the alarming signal throughout its pregnancy without ever needing to shock it again. Also we can now subject the pregnant female to greatly heightened anxiety by preventing its making the avoidance response which it has learned to make after this signal. This can be achieved by shutting the door between the two compartments, so that it cannot move to the other side when it receives the signal which used to herald the shock. Under these circumstances it can be shown that the young which it was carrying at the time will themselves show evidence of the treatment in later life. This is done by comparing such animals with ones which come from mothers not treated in this way. The latter serve as the essential 'control' group in any such experiment, without which it would not be possible to draw conclusions. The offspring, judged by the amount of activity exhibited, show greater timidity in exploring an open space. This suggests that the subsequent behaviour of these young animals has in some way been modified by the experience of being nurtured in the womb of a mother who was frequently made anxious. Moreover, the extent to which the offspring, which, though displaying no physical abnormalities, are nevertheless affected in their adult behaviour in this way, is to some extent dependent both on the intensity of the stress applied to their mother when she was pregnant, and on the stage during the pregnancy when it occurred. This has been shown by the use of doses of drugs which stimulate the release of glandular hormones of the same sort that are produced during the behaviourally induced anxiety discussed above. In addition, whether or not the offspring are affected is also characteristic of the type of mothering they have had after birth. This can be tested by the procedure known as fostering – giving the newborn rat pups to a different female to suckle until weaning. Such a foster mother can be one which had been treated in the same way as the mother herself, and stressed during pregnancy, or it can be one from the non-stressed control group. The effects of such a change can be detected in either case; sometimes they decrease the severity of the consequences of the pre-natal stress on the natural mother, and sometimes they increase it. The nature of the influences at work is still the subject of research and interpretation, but it seems possible that hormonal disturbances operating via the maternal blood supply to the developing foetus may be responsible. Here, perhaps, is a case in which the old wives' tales of dire effects resulting from pre-natal influences, often discounted, must be accorded some respect. The acquired effects may be small and subtle, and it takes careful behavioural observation to detect them, but the latest experimental work clearly suggests that they do exist.

The period immediately after birth is one which is important for the later expression of behaviour observed in adult life. It is well known that psychoanalysis places great importance upon the early development of social relations, especially with the parents, in young children as a factor in the way their person-ality develops in later life. While there is no comparable study of the way animals behave towards their young, and the re-sponse of the offspring to them (and indeed it is difficult to plan such experiments), nevertheless there has been much research in recent years on the role of early experience and its effects in the later life of animals. Thus it has been shown that the way baby rats are treated in infancy can affect the speed with which they learn when adult.

What is done is this. The young animals – 'pups' they are called – are taken from the mother rat from time to time during the period before weaning, which is at about three weeks of age in this species. They are subjected to various types of stimulation, stroking, petting, or even shocking by a mild electric current. The exact nature of the stimulation seems to be less important than the fact that it is applied, and even simply removing the pups from the nest once a day and replac-ing them will serve, for pups treated in this way show striking differences from animals which have been left undisturbed with their mothers. They show more advanced development in many ways: their eyes, closed at birth in rodents, open sooner, their weight is greater, and their hair appears earlier. If you do the same sort of thing with Siamese kittens, the adult type of coloration, including the dark feet, as opposed to the uniform off-white of infancy, develops sooner. All this bespeaks an acceleration of physiological development to a remarkable degree. Morever the acquisition of learned behaviour in adult life is affected, as can be seen in the speed with which the avoidance response in the escape-avoidance conditioning type of apparatus is acquired. Comparison is made with control animals who have come from the same or comparable mothers but who were not stimulated in infancy by one of the various ways mentioned. In this way it is possible to show that the group given infantile stimulation learn the response to the buzzer of crossing over to the other side of the apparatus more slowly than does the control group. Since the two groups have been otherwise treated in identical fashion except for the single factor of the stimulation, it is reasonable to assume that this is responsible for the difference observed. It seems likely that the infantile stimulation has reduced the tendency to be alarmed by the shock, so that the shock is less effective in motivating learning. These are antecedent influences which have operated during the past development of the individual – not, it may be mentioned parenthetically, during the more remote past and during the evolution of the species, as is the case with innate responses.

What is the role of environmental influences operating at the time the animal acquires the behaviour in question? What are the special conditions necessary for learning to take place? The most important of these is motivation. In Chapter 3, the primary importance of the motivation or drive in promoting animal learning was shown. In the laboratory this is easily observed and the effect of varying the nature and extent, for example, of the motivation has been studied. Drives can be broadly divided into negative and positive. Examples of both have been given: the use of food to induce the hungry rat to run a maze and the use of shock to cause it to escape to the other side of the conditioning box. Others have already been mentioned, and their relative effectiveness was the subject of much research during the 1930s. However, it is difficult to change the motivation and keep everything else in the experiment constant, or at least under control, which is what is required if one is effectively to compare the relative potentiating properties. Moreover, the question has lost contemporary interest. Students of animal behaviour are nowadays less concerned with knowing whether or not sex is a more potent attraction for female rats than hunger, or vice versa, just as they have come to realize the relatively unprofitable nature of speculations – even those based on experimental evidence – on which species is more 'intelligent' than others. The study of motivation is today concentrated more on the functional properties of drives which impel behaviour.

The way this is done in the laboratory is to manipulate the conditions which determine the intensity of the drive in the individual animal and then compare the effects of such differences upon some performance of interest. Usually several groups are used: one group might, for example, be made more hungry than another, which in turn was more hungry than a third. This can readily be done by manipulating their daily feeding schedule, so that the time elapsing between feeding and behavioural testing is different for the various groups. If instead of positive motivation, e.g. hunger, which is rewarded by food, a negative motivation, e.g. escape from electric shock to the feet is used, then the desired variation in intensity is obtained by varying the strength of the current allowed to pass. Alternatively the two methods may be combined, as in the air-deprivation technique whereby the animal is detained under water for a short time before being released to swim, reach the exit, surface, and resume normal breathing. Here again, the intensity of the motivation may be readily manipulated by simply varying the lengths of time of the initial delay, so that the longer you keep them under before they are released to swim to the exit, the faster they swim. But the interesting thing is that this is true only up to a point: there is an optimum delay which, if exceeded, causes not faster swimming, but slower, and so the rats' swimming speed goes down, and not up. This sort of relationship between motivational intensity and performance is a 'curvilinear' one, because it shows a peak in the middle of the graph. This curvilinear relationship probably applies to many sorts of motivation, though it has not been demonstrated in this way for all. It has its counterpart in our subjective experience in everyday life, and, though common sense is by no means an infallible guide to scientific facts, it does seem in this case that the result of trying harder to do a task is not always followed by an improvement in the way the job is done – it is possible to try *too* hard, with a consequent loss of efficiency. Now this finding about motivation and performance has had a further extension in the laboratory in the discovery that the point at which increase in intensity of motivation no longer pays off with improved performance depends upon the type of task being done. The easier it is, the longer the increases of motivation improve matters. Make the task more difficult, and the change-over to a decrease occurs sooner. Make it very difficult, and it is hard to detect any increase at all, increasing motivation merely serving to decrease performance right from the start. Let us consider an illustrative case. If, as before, we use rats swimming underwater under various degrees of air deprivation, then we can arrange matters in such a way that they have to learn to choose one of two alternative escape channels underwater, one being brilliantly lit and the other dimly. Now this they can readily learn to do – it is an easy task for them – and increases in air deprivation, as indicated by time of delay before release, cause improved performance. But if we make the dim light brighter, the difference between the two escape routes is lessened and the task is in consequence a more difficult one. Then we find that the change-over point from improvement to worsening the performance comes earlier, at a lower degree of air deprivation than before. If the task is made even more difficult by a further brightening of the dim light so that the two escape routes look even more alike, then the optimum delay is reached even sooner. So now we can say that the performance depends not only on the intensity of motivation, but upon the difficulty of the task set as well. This principle seems to have a wide application in many aspects of behaviour, including that of humans. Again an appeal to common sense may be permissible here: most people would agree that, in their experience, for doing something difficult and finicky like threading a needle, one shouldn't be too eager; if you are it tends to delay you – more haste, less speed – whereas for doing something simple and straightforward, like running to catch a bus, the more motivated you are, the more likely you are to succeed – other things being equal, of course. This leads to further interesting speculations. Perhaps for the extremely difficult intellectual pursuits which academics indulge in, the so-called ivory tower, which is thought of as devoid of the stresses and strains of the hurly-burly everyday world, is the best place. But this takes us too far from our theme, and the laboratory.

Another way of altering the motivation of the laboratory animal is not to try to influence its drive state by manipulating its feeding or its breathing, but to alter the size of the reward given. Thus within limits the speed with which uniformly hungry rats will run to a food incentive will depend on the size of the reward itself, the larger the faster, as you might expect. But alter the accustomed size, and some interesting things happen. Thus, a group of rats which had been receiving, say, a quarter of a gramme of food after each run will increase their speed if the reward is doubled to half a gramme each time. But the new speed will be faster than that of the group which had been receiving half a gramme all the time, though it may return to about this level later. Similarly, cutting the size of the reward by a half will have a slowing-down effect, but a group so treated will have a slower speed, at least at first, than the group given the same small amount all the time. These effects, which have been termed the elation and the depression effects respectively, have obvious analogy with similar ones known to exist from the studies of working output in humans, and with certain gambling situations. It is as if the rats were pleased by their increase in 'wages' in the one case and disappointed in the other by the cut, and responded accordingly, by altering their performance to suit. But this obviously is not a satisfactory explanation and smacks of anthropomorphism. Moreover, it suggests that the rats were in some way trying to pay out the experimenter in his own kind, which presupposes a degree of awareness of the situation on their part which they undoubtedly do not possess, and the presence of a relationship between the lowly rodent and his laboratory master which it is difficult to imagine existing. For example, there is no evidence that laboratory rats can distinguish one human from another in the way a dog can.

But with monkeys the situation is somewhat different: not only can they recognize people, but they can also be trained to accept payment for their services in a form of 'money' which can be 'spent' later. What was done was this. Instead of giving the monkeys a few raisins or peanuts for a reward in the type of discrimination experiment described earlier, the experimenter gave them instead a plastic token. These could be accumulated, and then placed into a slot machine which delivered grapes. The monkeys soon learned to work readily to obtain and amass these tokens. It is also possible to get monkeys to work for no reward at all – or so it seems until one looks into the conditions involved. In a Skinner box, they will press a lever constantly in order to operate a mechanism which will open a little window through which they can peep out at what is going on outside, and they will also spend much time on solving wire puzzles of the sort made up by door fasteners of the clasp-and-hasp variety. These findings suggest an exploratory or social 'drive' in monkeys which is providing the incentive for the concentration of effort which takes place. By the methods now familiar it can be shown that a genuine drive is operative in such situations. This is shown by the fact that it is possible for new learning to be established with such incentives as the sole reward for it. Thus, in the example given above, the monkey will learn a new task, such as lever pressing, to obtain the view out of the box.

Yet another aspect of motivation for acquired learning, which has given rise to much experimental work in the laboratory, has been the question of double drives operating together. Do hunger and sex drives, for instance, add up to create a super-drive which will make learning more effective than that based on either alone? There are several special difficulties about investigating this type of problem. It is next to impossible to find a learning task in which this proposition can be satisfactorily tested, for the reason that many drives involve others, at least minimally. So that a comparison of rats trained under, say, hunger drive with a similar group trained under both hunger *and* thirst drives is probably invalid, because making rats hungry causes them to restrict their intake of water, and vice versa. Thus the addition of the thirst drive to the motivational complex will make no difference, and so it becomes difficult to tell whether or not the two do 'summate'. The solution is to select two drives which differ from each other as much as possible in their physiological nature, such as escape from water (surface swimming) combined with hunger (food deprivation) or pain (escape from mild shock to the feet). When this is done, we can usually detect that the two together have a combined influence which is greater than that of either alone, in that the rat completes its task of swimming through a maze faster.

Another important aspect of learning which has been much studied in animals in the laboratory is the question of practice. What kind of practice is the most efficient for the rapid learning of a task, and what least? One idea which has emerged clearly from experimental work with animals is that of a learning set. This implies that if there is a predisposition to do something, then previous experience of doing it will facilitate its occurrence again. Thus, applied to learning new discriminations in the test apparatus for monkeys described earlier, it was found that new problems did not take as long at the end of a series of problems as they had at the beginning. There was a 'learning to learn' process at work.

Similarly, there has been an accumulation of evidence, mainly from studies of learning in humans, that spaced practice is more efficient than massed practice. That is to say, if you are to spend, say, half an hour on learning a short poem, it is better to spend the time in short bursts of practice with gaps between than it is to devote the whole half hour to continuous work. Animal study has in general confirmed this view, and has perhaps contributed rather more to the study of this type of question in offering conditions of learning which it would not be possible to employ with human subjects. Thus the nature of 'one-trial' learning has been investigated in the laboratory using rats by hiding an electrode which delivered a painful electric shock upon contact in a pile of sawdust where they had been trained to dig to receive food. The resulting aversion from the place was complete, and its nature – how long it persisted, what could be done to break it down, etc. – could then be studied. An analogous situation, in which strong punishment which could not be given to human subjects was used, employed avoidance conditioning of a sort in which dogs were required to leap over a low barrier in order to escape a rather strong electric shock. This procedure is not unlike that used with rats and other smaller animals in the escape-avoidance conditioning procedure which we have mentioned several times before. But, for the dog, having to jump over a hurdle provides a more suitable situation than having to move into another compartment.

Such habits, once learned, can be very persistent; the animal will go on making the response, time after time, for many months. And this brings us to the topic of memory and forgetting in animals, as it can be studied in the laboratory. Strictly speaking, forgetting is hardly 'acquired' behaviour, but rather the reverse, since it can be regarded as the process of 'dis-acquiring' habits. As we saw in our discussion of some kinds of animal behaviour studied in the field, there are well-authenticated examples of apparently long-retained memories of places and other individual animals. But in the laboratory, the concern has been less to see what records of longevity of memory the rat or other animal can establish, but rather to investigate the process of forgetting in order to examine its nature in man.

The technique which has been most widely used is that of the 'delayed reaction'. In this, the animal under test is shown the lure or bait, which is equivalent to the food reward in maze learning. It is then concealed from the animal's sight and the time elapsing before it forgets the location can be measured by restraining it for various intervals before releasing it to run to the hidden food. This was the older method of tackling the problem, and it is open to several objections. For animals larger than the rat, a large amount of space is required. Also there is some doubt whether we are here dealing with memory in the sense in which we understand it, that is, as a mental re-enactment of the experience which one is trying to recall, irrespective of the relative positions of the object and the place where the re-call takes place. Or is it a result of a bodily posture taken by the animal when the food is first hidden, and which is used to 'point' to the correct place during the waiting interval? In this way the animal's behaviour could seem to be the result of a higher mental process than it actually is, and we would be deceiving ourselves if we were thereupon to attribute to the monkey the power of true memory, which it perhaps does not possess. This is an example of the sort of difficulty mentioned in Chapter 1 which we encounter in the interpretation of animal experiments, and the solution in this case is obvious. We must ensure that the animal does not make use of the clues it can derive from pointing and maintaining a bodily posture in the way described. This is done by anaesthetizing the animal in the interval, or requiring it to make some bodily movements in order to destroy the maintenance of any bodily posture or 'pointing'.

The concept of the memory 'trace' – which it will be remembered is only a shorthand way of referring to the bio-chemical changes which probably take place in the brain tissue as a result of learning, the nature of which is not yet fully understood – can conveniently be investigated in animals. At what stage is this memory trace consolidated after the experience which gives rise to it? This can be investigated experimentally by allowing rats to learn a maze, all to an equal degree, and then giving some a convulsion like an epileptic fit, which can be induced by passing an electric current through the brain. The procedure is quite painless, the electrodes being attached to the outside of the head and the animal losing con-sciousness immediately. Now this procedure is known to induce forgetting of what preceded it – retrograde amnesia, as it is called. Hence, by giving the shock at various times after the training in the maze, and then comparing the different groups who had it at these different times in respect of the amount of the training they retained (this is done by testing all the rats on the same maze at a fixed time later), we have a situation in which all animals have been treated the same way as regards training, testing, and inducing a convulsion. The only thing that varied was the place in this temporal sequence at which the convulsion occurred. Experiments along these lines show that the convulsion must occur within one hour of the initial learning for the disturbance in the consolidation process to be detectable in the second test. This suggests that there is a need for an undisturbed period of time to pass for learning to be most effective and the memory trace to be consolidated.

Allied to the study of the delayed reaction is the measurement of time discrimination in animals. How well can rats, for example, estimate the passage of time? The answer, surprisingly enough, is 'very well'! The way it is done is this: Imagine a maze with two paths to food; it could very well be T-shaped, so that the rat will learn to run along the path represented by the upright of the letter, then turn either left or right, and either way receive its food reward. Now, other things being equal, you would expect half the turns at the choice point to be to the left and half to be to the right, and this would still be so if you were to arrange for the rat to be delayed before receiving the food so long as the times on the two sides were equal. But make them unequal, so that turning to the right meant waiting, say, one minute, before the food was given, and turning left meant waiting two minutes. Now we should expect a preponderance of right turns if the rat could make the distinction between one and two minutes. In this way we can try two different times which are closer together until we can see by the way the animals behave that they can no longer tell the difference and discriminate between them. In experiments of this sort it has been shown that rats can distinguish between such short periods as ten seconds and twenty seconds with a fair degree of accuracy. An alternative way of proceeding in the measurement of the time estimation of the rat is to use the Skinner box described earlier and arrange to have a delay associated with the lever, just as in the alley maze mentioned, by altering the mechanism responsible for delivering the pellet of food so that it only works after a delay of so many seconds after it was pressed. The delays can then be varied much as before, and we can tell by the rat's speed of responding to the bar, when it has a relatively short delay associated with it, that the animal can discriminate this time-span. Then we can make the delays longer and longer until the rat shows by the absence of any learning that it can no longer associate the pressing of the lever with the delivery of the food reward and has reached the limits of its powers of time estimation. Using this method has yielded somewhat different findings regarding the time estimation capacity of the rat – it can learn a ten seconds delay but not a thirty seconds one. This discrepancy is typical of the sort of differences which different methods bring out in the study of animal behaviour. The strain of animals used often varies from experiment to experiment, and this also can give rise to differences in the outcomes.

Pigeons working in a Skinner box have been shown to be able to discriminate time. The experiment is typical of the many ingenious arrangements which can be used for studying the behaviour of birds in this kind of apparatus. What was done here was to arrange for a lighted key, which the birds had been trained to peck at for a food reward, to go dark for various periods of time. They then had to learn to cease pecking after any but the shortest of the various periods of darkness, since they were no longer given food except after this one. The change in the rate of pecking showed quite clearly that they could detect a difference of three seconds out of thirty.

Use of the Skinner box in yet another connexion again confirms the sensitivity of the rat to the passage of time. Among many possible patterns of reward and no-reward following each pressing of the bar, one consists of the delivery of a food pellet by the mechanism at a fixed time interval, completely irrespective of the number of times the bar is pressed during that interval. It is found that the frequency of bar-pressing by the rat increases sharply as the time approaches for the reward to occur – a sure sign of awareness that the time interval is approaching its end.

The ability to learn to estimate time is quite striking among lower animals. Can the same be said of their ability to discriminate number – that is, to count? Now it is quite clear that animals can be readily taught to discriminate one pattern or shape from another, and to demonstrate that they have acquired such power by behaving differently according to which is shown them. This we saw in our discussion of the salivary conditioning procedure in dogs, originated by Pavlov. The patterns might consist of different numbers of, say, black dots on a white background, so that we might be tempted to conclude that the animal was able to count if it could tell four dots from five. But this would be an improper deduction from the experimental observation, for it would almost certainly be the case that the animal was responding not to the absolute number of dots, but to the configuration presented by the four dots, which must, of necessity, be different from that made by five, although in such a slight way that it may seem near to miraculous to us that the animal can detect it. Indeed many of the apparently remarkable achievements of the animal world, for example, navigation by birds or the recognition of individual mates or offspring in large colonies numbering many thousands, are probably based on the acute perception by the animal of minute differences which humans either could not detect except perhaps after very long practice, or cannot detect at all because they do not possess sense organs acute enough or of the appropriate kind.

But to return to the problem of counting by animals. With-out assuming its existence on doubtful grounds, we find that the most successful methods have involved a combination of delayed responses and shape recognition. For example, a bird is shown a pattern of five pieces of material on a disk. It then has to find, from among a series of such disks, the one which also has five marks on it. The important thing is that the marks are all irregularly shaped and are different on the original test disk and on the one which it matches in terms of number. Using blobs of plasticine to make the various numbers, and training magpies to push aside a disk to reveal a piece of food in a pot beneath, it has been shown that some birds could recognize numbers up to eight formed in this way, and demonstrate their recognition of them by going over to the pot with the appropri-ate number marked on it in plasticine pieces and choosing it without error from among many others.

Another question relating to the acquisition of behaviour, which has been much studied, is whether certain species of animals can learn by imitation of other individuals of the same or different species. The spread across Britain of the habit of several species of birds, but especially tits of various kinds, of raiding the milk bottles left on doorsteps may have been caused in part by imitation. The birds will raid the bottles, puncture the cover, and drink the cream on top of the milk! The development of a similar sort of cultural habit has been observed among the monkeys on the Japanese island referred to in Chapter 4. A female monkey started washing her food – sweet potatoes – before eating it. In three years the habit had spread to eleven other individuals on the island. Experiments have shown that rats can learn to imitate trained rats in learning to navigate their way through a maze. What was done was this: A rat that had been trained to run through the maze perfectly was allowed to start immediately ahead of a novice – the 'naïve' animal – which was then rewarded with food according to whether or not it followed the first rat. Rats which had this guidance succeeded in learning the correct path through the maze faster than a second or 'control' group which did not have the benefit of the guidance. A similar finding has been made from training rats in avoidance conditioning in the presence of other rats which were already expert in the task of responding to the buzzer, having previously had the usual training in learning to associate the buzzer with forthcoming shock and to avoid it. But the trainees were never shocked; they were merely exposed to the experience of being cooped up in the shuttle box with rats which ran to the other side whenever the buzzer sounded. When later tested alone, however, they responded to the buzzer as if they had been through the usual shock-escape training routine.

Cooperation between different species of animals can also be induced, but it smacks of a circus trick such as the trained dogs riding on the back of the trained horse. The very real skill and profound understanding of the principles of learning displayed by many animal trainers should not be despised, but it hardly adds anything to our scientific understanding of the processes of behaviour involved. However, there does exist experimental work which illustrates the role played by learning in the apparently innate antagonisms between various species – especially those which normally prey on one another. Thus it is possible to train those apparently natural enemies the cat and the rat to cooperate, when it is to the advantage of the predator (the cat) to do so. This has been done as follows. The access to food was barred by a door which could be raised by stepping on two pedals simultaneously. These pedals were too far apart for the cat to reach at once. Now a rat which had privately, so to speak, been trained to step on a pedal to raise just such a door was introduced, and the cat soon learned to tolerate the rat's presence when the rewarding consequences became apparent, that is, the speedy opening of the door which was not otherwise possible. But we are not told what the consequences might have been had the cat in question preferred the live rat meat in front of the door to that provided by the experimenter behind it!

The effect of age on behavioural processes was touched upon in the previous chapter in discussing the extent to which the innate processes there described are affected by age. In the same way, acquired learning is affected by age, and there is good evidence to suggest that the actual acquisition of new habits is impeded. Here again there is evidence from experimental work with humans of various ages, but the convenience of animals for this kind of study has not been overlooked. The life span of the rat, for example, is quite short, so that it becomes possible to control external influences other than age relatively easily, in a way not possible with human studies. A three-year-old rat is old indeed, for it may be generally taken that one rat year equals thirty human ones. Thus by judicious manipulation of breeding dates over a relatively short period of time, a population of animals can be obtained in which any or all the desired ages are represented. In investigations of this sort, in which the capacity of the animal to assimilate new learning is studied by a maze-learning technique, it was found that rats over two years of age showed a slower learning of a new maze, but there was little relationship between the ability to remember a maze previously learned and any visible signs of senility. Thus, old age in animals shares some of the characteristics of old age in humans, at least in the intellectual sphere. A failing memory and an inability to remember new names and faces have their counterpart at the animal level.

Before leaving the topic of acquired behaviour, there is one allied subject which is of some interest. This is the doctrine of the inheritance of acquired characteristics, which is primarily associated with the names of Lamarck and, more recently, of Lysenko. At its crudest level it can be expressed as the expectation that if you cut off the tail of, say, generation after generation of mice before breeding them, you would in time find yourself with a strain of Manx mice. Now this seems inherently unlikely on the face of it, but there are more subtle variations of the theory which have greater credibility, and indeed there are complex genetical situations which appear to give rise to results which seem to suggest that characteristics have been acquired and have begun to act as part of the hereditary make-up of the organism being studied. There has been little serious attempt to adapt this doctrine to learned behaviour, which is, after all, an acquired characteristic *par excellence*. One notable series of experiments was performed, which, together with the equally notable series of experiments undertaken to refute the results of the first set, caused some stir. What was done was this: Rats were trained to escape from a tank of water on to a dry platform by one of two routes. One of these was brightly lit, but was associated with a grid which gave a painful shock to the feet; the other was in darkness and was safe from this hazard. The animals learned naturally enough, in the nine daily trials they were each given in the tank, to avoid the shock and to choose the darker escape route. Now, in successive generations, there was a definite tendency for the offspring to learn this habit with fewer and fewer errors. That is, there appeared to be an inheritance from one generation to the next of an acquired characteristic – in this case a facilitation due to the effect of the training. This was so contrary to what might be expected on general grounds that at least two repetitions of this work were undertaken, notwithstanding the fact that such experiments are inevitably time-consuming and tedious, because of the necessity to go on year after year, with successive generations. While the rat breeds quite quickly – pregnancy lasts twenty-two days only – the young have to grow to adulthood before they can be tested and bred for the next generation, so that only about three generations can be raised each year. The initial experiment mentioned employed thirty-eight generations, and the most complete among the repetitions employed fifty generations and took twenty years to complete! The general findings in this repetition of the experiment were negative; it was shown that the appearance of a progressive increase in the speed of the acquisition of the escape habit could have come about through fluctuations in the general health of the colony, and also that there may have been a selective influence upon those animals chosen for breeding. This could, of course, give rise to mimicry of an inherited effect, for as we have seen it is well known that selection for a particular behavioural characteristic is effective. Here animals, deliberately chosen for their possession of certain genetical characteristics, are bred together, all others being excluded from the breeding for the next generation. This is based on the proven expectation that doing so will increase the representation of their heredity in the general population, and so there will be a rise in the average level of the characteristic bred for. This of course is a different matter from the onward transmission of characteristics actually acquired by the individual during its lifetime, and so the position at the moment is that experts discount the possibility of such inheritance.

6

The Abnormal Behaviour of Animals

OUR knowledge of abnormal behaviour in animals depends very largely on our knowledge of abnormal behaviour in humans, because it is only by reference to human abnormal behaviour that we can evaluate what we observe in animals. The reason is that there are, with very few exceptions (such as the dog and some other domesticated animals), no animals of whose normal behaviour we have so intimate a knowledge that we can pinpoint a change in it and say that it is now abnormal. Also the line between sanity and insanity in humans is by no means the clear, cut-and-dried affair that is often imagined. Behaviour which would be regarded as insane in one society or civilization might be normal in another, and for scientific purposes the best criterion of insanity in our society is to regard the line as having been passed when it becomes necessary to send the patient to hospital, either for his own protection or that of others. But we do not send animals to hospital for neurosis, and so this will not do for them. The next best thing, therefore, is to try to define what we mean by abnormality – or at least neurotic abnormality – in humans and then see if we can supply this as a criterion to animals. Here we run into another difficulty, the absence of speech in animals. Now this is nothing new, and we have had many examples in this book of getting round it in the study of animal behaviour. But in the present matter it is more serious, because human abnormality is so often detected by what the patient *says*. He may complain of extreme depression, of feeling that he is changing in some way, and so on. We can counter this by defining human neurotic disorder in behavioural terms, that is merely by reference to what the typical patient does, not what he says. We then have a behavioural yardstick by which we can evaluate what are thought to be similar cases in animals. But let us consider some such cases and see what the behaviour is that has given rise to the term 'experimental neurosis' in animals.

The name is due to Pavlov, who first detected abnormal behaviour in dogs undergoing his salivary conditioning. He found that it occurred principally in the situation in which the dog was being called upon to distinguish two different stimuli, in this case a shape discrimination between a circle and an ellipse. It will be remembered that this is done by making one of the stimuli, here the circle, the positive stimulus and always following its appearance with food, and the other, the ellipse, the negative one, and never rewarding the dog with food after it. Now Pavlov was interested in seeing how small a difference between the circle and the ellipse the dog could recognize, and so he made the ellipse more and more like the circle on successive occasions. This went on fine for a time, and the dog learned to recognize the difference between the two when the ellipse was nearly circular. But the attempt to get a more precise discrimination by making the ellipse even more like the circle led to a strange happening. The previously placid dog now reacted violently to the procedure, and fought the restraint of the harness in which it was confined. It howled and resisted being brought to the laboratory. When it could be induced there it showed that it could no longer respond to even the simplest discrimination of a circle from a grossly elliptical shape, which it had done readily prior to the 'break-down'. Pavlov termed this concatenation of behaviours 'experimental neurosis', and went on to devote much research to it. American workers took up the study in the period before the Second World War. A variety of different species of animals was studied, and several different methods of inducing such abnormal behaviour applied. Common to all of them, however, was some sort of 'conflict' between opposite tendencies created in the animal by training methods, just as in Pavlov's original work. Thus, sheep were trained to bend a leg in response to one signal and to refrain from doing so when a different one was shown, and in this way difficult discrimination problems were arranged for them, just as with Pavlov's dogs. The same symptoms of restlessness, refusal to come to the laboratory, and loss of previously demonstrated powers of discrimination were all seen. Some sheep developed a sort of insomnia, so that instead of sleeping at night they spent the time restlessly pacing around. Previously placid animals became aggressive, and some started grinding their teeth in a tense sort of way. With dogs there was a similar story, and one dog, the celebrated 'Nick', was studied over a period of no less than twelve years. He first started showing disturbed behaviour in a conditioning situation not unlike that used by Pavlov. In addition to refusing to respond to the test signals and to eat the food provided in connexion with them, he developed other apparently abnormal symptoms. A loud, raucous breathing was noted, and there were inappropriate sexual reactions – penile erections without the presence of the normal stimulus of a bitch in heat. Furthermore, the dog behaved in these ways even when it was given rests from experimentation and was holidaying on a farm in the country away from the laboratory.

Pigs and cats have also been used in this work; they were trained in slightly different fashion. Food was put into a small trough provided with a lid, and the animal had to open the lid in order to obtain food when a particular signal was given. If it opened the box at any other time, it received no food. In some cases the situation was made worse for the animal by scaring it by a blast of air, or electric shock to the feet, or both, just at the moment it was about to feed from the box. The cats, especially, reacted violently to this treatment. They refused to come to the experimental room, refused to respond to the signals announcing food in the apparatus, and refused to eat it when provided. Often they remained motionless near the food box. Indeed, with one pig this behaviour was so marked that an apple – the animal's favourite food – could be balanced upon its nose and still it would not eat. All these aberrations of behaviour more or less resembled what Pavlov had called 'experimental neurosis', and were regarded as neurotic by the workers concerned.

But what happens if we look at these examples of abnormal animal behaviour when considered by the yardstick of human neurosis, defined in behavioural terms, without the intervention of speech in the way we considered earlier? In some ways they resemble what we regard as essential components of neurotic behaviour in humans: emotion is disordered, the animals show evidence of conflict, and these manifestations go on for a long time and show in many ways in the individual's behaviour. But human neuroses are shown by only a few people in any population – the way most people behave is the norm by definition, and only a small proportion of the population can be regarded as neurotic. An even smaller proportion are actually put in hospital, and this, it will be remembered, is our criterion of neurosis. Also, it can be argued that the definition of what is normal has another connotation – the evaluative. That is to say, the behaviour that is clearly bad for the individual, or for the species as a whole, can also be considered biologically abnormal.

Now, when we look again at the examples of animal behaviour which have been thought to be equivalent to human neurosis in the light of these criteria, what do we find? There really are very few cases reported which come anywhere near meeting them. In all cases we can see evidence of emotional upset, and this is not to be wondered at in most cases, considering the unpleasant nature of the experience which the animals are necessarily subjected to in this type of experiment. This is biologically normal, and indeed it would be a bad thing for the species as a whole if individuals did not seek to avoid danger and frustration. It is not possible to consider behaviour shown by an animal in a threatening situation as abnormal – that is, when the animal does something apparently stupid which does not solve the situation for it – without presupposing powers of deduction that the animal almost certainly does not possess. That is to say, if the animals had been allowed to do so they might have starved rather than face the shock or threat of it. Yet we cannot regard this as biologically abnormal, unless we assume the animal really knew that the shock itself was not potentially fatal, that the experimenter was not, in fact, trying to kill it. So the refusal to feed can be regarded as a reasonable response to danger in the circumstances, and one which might be best both for the individual and for the species in the long run.

As already noted the emotional behaviour often persists for a long time in the animal, and this resembles human neurosis, but in very few cases is there evidence that it is shown outside the situation which provokes it, and this is unlike human maladjustment. One such case is that of the dog Nick, mentioned above, which displayed its symptoms even when in the country. But it should be noted that they were only observed when people from the laboratory came to visit it, and so, though it may seem that this dog was taking its troubles with it into the country, it was perhaps the case that this was the result of following it there with stimuli in the shape of people from the laboratory. As for there being any significance in the sexual abnormalities observed in this dog, this is open to grave doubt, since this animal had been previously used in an experiment in which an attempt had been made to condition sexual arousal to various stimuli deliberately, just as salivation can be conditioned to bells or lights. It is therefore very probable that an association had accidentally been built up between sexual arousal and the sight, etc., of the research workers taking part in the experiment. This would be an example of a normal, if somewhat unusual, form of learning by conditioning, and need not have any abnormal connotation at all. Similarly the evidence from these animal studies that the behaviour observed was abnormal in the sense that it was displayed by only a minority of the population, as is the case with human neurosis, is completely lacking. This has never been demonstrated in dogs, the species most studied, nor indeed in any other that has come under observation. So it appears that the notion of an experimental neurosis which can be created in animals in the laboratory is not a very satisfactory one to work with. Such a conclusion is unfortunate for progress in the study of human mental disorders, for it means that there is no standard animal analogue of the human disease which can be used for intensive study and experimentation in the way that, for example, cancer can be deliberately induced in rats and mice for laboratory studies of its onset and possible cure.

This does not mean that the scientific study of animal behaviour has nothing to offer in the way of a greater understanding of human mental aberrations. There are numerous indications from both field and laboratory work that valuable insights can be gained from animal study. We have already seen (in Chapter 4) a striking example in the laboratory work on the disorganization of behaviour in the rhesus monkey occasioned by the deprivation of maternal care and affection in infancy. But it is clear that there is nothing which has yet been identified directly akin to human neurosis and psychosis in animal work, and which can be used for the purpose of study. This is not surprising for those who take the view that the study of normal behaviour in humans is the key to understanding the abnormal, since the normal almost imperceptibly grades into the abnormal. As has been stressed before, there is no hard and fast line of insanity; rather it is a matter of degree and of the culture in which the behaviour is being observed. Therefore the study of the normal illuminates the abnormal, which can be viewed as the outcome of the same forces, both environmental and constitutional, as those which determined normal behaviour but which are perhaps more intense both in action and outcome.

Of the apparently abnormal behaviour, which may be observed in the wild, the periodic mass migrations of lemmings, in Scandinavia especially, are the most dramatic. These little animals, related to the mouse and the mole, are normally shy and retiring, and are rarely seen during the day. They live in the upper valleys in the northern mountains. But every four years or so, in different localities, an increase in their numbers occurs, they denude their normal ranges of the plants upon which they rely for food, and then they set out in hordes down the valley towards the sea. During this time they appear quite bold and will even defy a man if cornered. They do much damage to crops in their path, but are greatly preyed upon by other mammals, foxes, weasels, etc., as well as by hunting birds, hawks and eagles and the like. The numbers of these predators also rises with the abundant food supply provided by the increased numbers of lemmings. Eventually, the movement reaches the sea, as it must with the downhill movement in northern Norway and Sweden, but this does not stop the lemmings. They plunge in and cross fiords or swim to offshore islands, many perishing in the process. Outbreaks of 'lemming fever' in humans may follow in their wake, probably owing to handling of the carcasses.

Now the reason for these periodic increases in population leading to the mass migrations are not properly understood. The indications are that they are the result of a natural increase in numbers from the low figures obtaining after the last population crash associated with a migration, partly owing to the absence of predators which have themselves become scarcer with the decrease in the lemming population. The migrations appear to take place as a result of the lack of food combined with a form of hormonal disturbance occasioned by the stress of overcrowding. And it is this latter which is the more interesting to the behavioural scientists than the apparently more dramatic and abnormal mass movement. The migration, after all, can be interpreted as being advantageous for the preservation of the species as a whole and so not abnormal in the broader biological sense – if all the lemmings tried to retain their original range the result might be total – if local – extinction. But the stress disease is itself of great interest, since it has affiliations with the sort of thing which is suffered by humans under various circumstances of danger or worry, and any increase in our knowledge of the way it comes about in the lemming is likely to help in the diagnosis and cure of the related human conditions.

Little is known of other instances of apparently abnormal behaviour in wild animals – rogue elephants and the like. So it is to the laboratory that we must turn for the further analysis of normal behaviour in the expectation that it will reveal the bases for the normal and hence the abnormal in humans. Here the study of the genesis of emotion is of crucial importance, in view of the incidence of emotional disorder in mental ill health. The discovery that emotional responses, just like other habits, e.g. the route to a goal in a maze, can be learned and unlearned, is of great importance.

What is involved in this type of learning can best be illustrated by an example. In the escape-avoidance conditioning apparatus discussed earlier, the rat or other small laboratory animal learns to respond to a signal which heralds the onset of an electric shock, and makes its escape without waiting for the shock itself to be turned on, which it used to do before the significance of the signal was learned. Now, what has happened here? It is possible to take the view that the animal has learned to fear the signal just as it previously feared the shock. On this view, therefore, just as an animal can learn a trick, it can also learn to feel anxious in a given situation. Now, fear or anxiety is a potent emotional motivation, as we have seen, and can be used to induce the performance of many tasks in animals in the laboratory. The question then arises, are we justified in thinking of the learned variety of fear as akin to the natural type which every animal would feel in the fear-arousing situation? The answer, broadly speaking, is 'Yes'. In a classic experiment, it was shown that rats which had been trained in escape-avoidance conditioning of the sort previously discussed would learn a new way of escaping from the signal alone, without the necessity ever to use the shock again. What was done was this: Their escape route was barred, but they were given access to a little switch in the wall which they had not encountered before. The function of this little switch was to raise the barrier. The rats soon learned to operate the switch, raise the barrier, and so make the escape to the 'safe' side of the apparatus.

Now this property of learned fear has been demonstrated in many other ways in laboratory experiments and undoubtedly is of great significance. And yet, as is often the way, it remained unexplored for over twenty years. Since the Second World War, however, the analysis of such learning of acquired drives has been pursued with some vigour. The greater our understanding of the way in which the animal analogues of anxiety as seen in the escape-avoidance conditioning experiment are learned, of the hereditary determinants which predispose the individual to learning them, and of the type of situation in which they are likely to occur, the greater in consequence is our ability to offer a rational treatment of human neurosis. For there is a school of thought which believes that human neurosis is formed on some such basis, that the origin of the misplaced emotional arousal so characteristic of this disorder in humans is to be found not in unconscious wishes, imperfectly repressed, but in the misplaced attachment of emotional responses to inappropriate signals, or situations in the case of humans, so that for the rest of the person's life he or she is responding, often with anxiety, to conditions which do not warrant it. Now this is clearly an over-simplification of what is certainly a difficult and intricate matter, and one which it will take many years to work out at all satisfactorily, but what is clear is that animal experimentation will permit a more exact specification of the many complex factors involved than will reliance on observation with humans alone. The point has often been made in this book that animal experimentation can be substituted with advantage for research into human processes, and here again is a case where this may be especially relevant.

Let us consider the work which has shown that rats and monkeys can be subjected to procedures which will give rise to stomach ulcers. Here is a case where the interaction of psychological or mental and physiological or bodily processes is especially close. It has long been accepted that psychological strain may give rise to stomach ulcers of various sorts in humans, though there are other contributory factors, of course. One hereditary one may be mentioned: it is known that people of certain blood groups are more disposed to contract ulcers than others. But it turns out that this type of disorder can be initiated in laboratory animals by processes which, on the face of it, look very like those thought to precipitate ulcers in humans. In the early work rats were used, and it was found that subjecting animals to severe threat of shock when they tried to feed, especially when they were rather hungry, led to the development of ulcers. These could be detected when the animal was dissected after death. And similarly with monkeys, but here interesting new facts came to light. In an arrangement in which a monkey was able to learn to press a lever in order to operate a mechanism which would delay a shock to the feet, it was found that animals which were denied this possibility of delaying the shock and received it every so often, or even at random intervals, did not develop ulcers, whereas those that had to concern themselves constantly about the need to keep the shock switched off did. These were termed the 'executive' monkeys, on the view that the sort of strain the procedure put upon them was not unlike that suffered by people in certain types of business.

Another kind of abnormal behaviour in humans which can be conveniently studied in animals is addiction. It has long been known that some animals, again especially monkeys and rats, can become addicted to certain drugs such as morphine and alcohol. Monkeys who have been experimentally addicted to morphine and then had the drug discontinued show classic 'withdrawal' symptoms; that is, their whole bodily functioning has become dependent upon the ingestion every day of a certain amount of the drug, and when this is withdrawn the readjustment is a painful process, just as it is in humans. The advantage of working with laboratory animals in this type of problem does not need stressing. New techniques and new cures can be tried out with greater rapidity and certainty than with a group of human patients suffering from addiction. Here again there seem to be predisposing genetic factors involved to some extent, and these can only be studied in the laboratory. Thus, while most rats and mice prefer plain water to water containing a proportion of alcohol, there are some which prefer the cocktail consistently. This is shown by arranging in each cage two sources of water supply instead of the usual single one, so that each animal can choose the drink from the bottle on the left or the one on the right. These bottles are carefully graduated so that each day it is easy to read how much has been drunk from each. In this way it is possible to determine the preference for whatever is in the bottles, merely by seeing which one has most consumed from it. It is necessary to switch the bottles around from time to time for it has been found that even if both contain exactly the same liquid, pure water for example, rats and mice will develop a preference for a certain position, one side or the other. There is then the danger, without the precaution of changing the bottles around, that what may seem to be a genuine preference for one or other substance may merely be a preference for one side over the other. Now, when such a preference has been established it is of great interest to discover what is determining it. One cause so far discovered is the hereditary one – some strains of mice will prefer alcohol whereas others dislike it, and this innately determined preference may be related to the addiction to the drug. Much work remains to be done in this connexion, as in many others mentioned above, but the advances made are encouraging and speak well for progress in surmounting some of the difficult problems in the understanding and care of mental illness.

7

Practical Uses of the Study of Animal Behaviour

THIS book has been concerned with the application of the contributions of the scientific study of animal behaviour to the understanding of human psychology. This is not to say that the study is not of great interest in its own right. It is, and there are many fascinating problems awaiting solution which have no present promise of being relevant to human behaviour, often because the processes involved are so different from those in humans. Thus the echolocation system used by bats as a sort of radar guide in hunting moths and other winged prey is of great interest, but in principle it is of little relevance to human communication, except perhaps for the blind, because the sounds the bats put out are above the range of human hearing (for all but children, who have a higher limit), and nobody can detect with anything like the bat's accuracy any sounds bounced off objects as they do. (The mechanism by which certain moths are able to detect the bats' 'radar', and thus have a warning signal when they are being hunted, is even more fascinating, and resembles the wartime device which served the same function, as well as the gadget which will enable the motorist to be warned of a police radar speed trap.) Nevertheless, science is interested in knowledge for its own sake, and the relevance to practical human applications of any problem is never, and should never be, the criterion of what is worth while. Even if it were the only criterion, it would be a rash person who would make a decision on the promise in this respect of a particular piece of research work. The history of science is full of examples of discoveries which seemed completely without any possible useful application at the time but which turned out to be of great value later. However, our view has laid stress in this book rather on applications to the problems of human behaviour, and the time has now come to sum up some of the credits and debits which we have sought to put before the reader.

It is clear that the final solution to any human behavioural problems can rarely come from animal studies; usually only hints, clues, and perhaps useful methods are the outcome. But these can be of crucial importance when it comes to trying out a new cure on human patients. Thus, in two striking examples of physical methods for alleviating the suffering of the mentally ill, electric-shock treatment and brain operations for certain types of disorder and depression, the suggestion for both came quite directly from observations upon animals. The electro-convulsive shock treatment as given to humans is essentially the same as that mentioned in Chapter 5. The convulsion, which is thought to be the important therapeutic factor in the treatment, is induced by the passage of a current through the patient's brain. Prior to the introduction of this electrotherapy, the convulsive effect had been obtained by the administration of more or less noxious drugs which had undesirable side effects. Now observations on animals in Italian slaughter houses, where the cattle were stunned by an electric shock before being actually killed, showed that a convulsion usually occurred as a result of this treatment. This suggested the use of electric shock in the human convulsive treatment, and it proved to be superior to the previous pharmaceutical method.

The observations which led to the introduction of brain surgery for the relief of certain kinds of nervous disorders were perhaps more systematic than is suggested by the account of the introduction of electrotherapy. What happened was this. In a study of the effects of removing part of the brain on the performance of difficult discrimination tasks by a chimpanzee, it was found that the emotional consequences of failure – outbursts of temper, refusal to work further (the sort of behaviour described in the previous chapter as characteristic of the so-called 'experimental neurosis') – were absent or greatly reduced. The animal was now placid and stolidly accepted its errors and their consequences. It was then found that the same effect could be achieved, not by removing portions of the brain, but merely by cutting certain nervous tracts within it – a delicate, though relatively simple, operation. This operation, which had its direct origin in animal experimentation, has been performed on many thousands of mental patients with much relief of suffering.

Both these techniques for cure have, however, been largely superseded by the advent of new drugs, and perhaps it is in this relatively new field of the study of drugs which influence behaviour in various ways that the utility of animal studies is most clearly seen. These drugs have produced a near revolution in our mental hospitals in the treatment of various sorts of mental illness, and many new ones have been introduced in the past few years. But before a drug is given even a small-scale trial on human beings there is a long and intricate process of testing to be done. Many different variants of the drug are tried, some of which differ very little from one another chemically, in order to find the best formula. Then its efficacy for the job it is designed to do is tested. Just as a drug designed to be effective against some forms of cancer would be tested on animals who had cancer experimentally induced in them, so a drug designed to counter, say, anxiety in human patients would be tested upon animals who were displaying anxious forms of behaviour. For this, the escape-avoidance conditioning experiment which we have mentioned several times might be used. And a possible method would be to compare two groups of rats, both of which had been taught to avoid the shock as soon as the buzzer signal sounded. One of them would be given the drug in a suitable dosage, and the other – the familiar 'control' group – would be given no drug at all, or perhaps a dummy form of the drug. Then, by comparing the performance of the two groups while the one was under the influence of the drug, the experimenters could tell if it had been effective in decreasing the anxiety aroused by the buzzer signal. But there is a snag here: it may be that the drug had merely reduced the efficiency of the animal in the way it hears the signal – a dulling of the senses rather than a dulling of the actual anxiety reaction. This problem could be taken care of by studying the way the escape-avoidance conditioning is learned in the first place, and seeing whether or not it is affected by the administration of the drug during the learning process. This complicates the experiment, of course, but increases our knowledge of the way the drug is acting, and, by comparing its action with that of other drugs, already known as potent in the relief of anxiety, we could tell more about the way it is obtaining its effects, if any. All this must be done before it is tried out as a cure for human suffering, and this screening, as it is called, can be a very lengthy process if there are many slightly different versions of a chemical formula to be tried out.

Animal observation and experimentation contribute to the understanding of human behaviour in other and less obvious ways than in the treatment of mental ill health. Many of these have been briefly referred to in previous chapters: the development of our understanding of the way the mind works in normal behaviour depends upon animal work largely for trying out promising hunches; the study of the evolution of behaviour, and especially that most human of activities, speech, can be furthered by animal work; and even the analysis of human patterns of forming and maintaining more-or-less civilized societies can benefit from the observations made on social animals and birds. But let us consider some of the ways in which a scientific knowledge of animal behaviour is of more direct use to us, either for good or evil, remembering always that science is neutral and the applications to which it is put are not necessarily those which were originally intended or even foreseen.

Animals have played a part in the exploration of space, and the Russian dog 'Laika' has been honoured for her part by at least one postage stamp. More recently a chimpanzee preceded the American astronauts' orbital flights by several months. In addition to the purely physical records of the reaction to the stresses of the space flight – the increased and decreased gravity, particularly – as measured by respiration, pulse rate, etc., the animal's continuing competence to execute conditioned responses was observed. The ape was required to press a bar every so often to delay a slight shock to the feet in order to evaluate its estimation of time, and also to make visual discriminations between various signals presented to it by means of flashing lights. Other small laboratory animals, including mice and rats, have been used for similar purposes in space rockets.

Animals have been recruited for warlike purposes, and while this may be deplored, it cannot be said that man has demanded of them more than he himself has, on occasion, been prepared to give. The Japanese Kamikaze (the name means 'wind of God'), the pilots who dived suicidally to their death on the decks of enemy naval vessels in the Pacific during the Second World War, were just as surely trained in their way by cultural pressures as were the Russian dogs which sought out enemy tanks to blow them up with explosive charges attached to themselves. It has been reported that during the Second World War seagulls were trained to assist in the detection of hostile submarines by releasing food from friendly ones, so that they would seek out and follow all craft of this kind and thus reveal their presence. Obviously this system would have its dangers in the absence of a satisfactory method of training the birds to discriminate enemy from friendly boats. But perhaps the most spectacular of all such employment of behavioural possibilities was the projected use of pigeons for visual guidance systems for flying bombs. As we saw earlier, the bar-pressing of the rat in the Skinner box, or the pecking at a spot by a bird, can be made extremely persistent, so that the animals will go on doing it for many hours in the absence of any reward, or perhaps with only a very minimal one. Now, a pigeon can clearly be trained to peck, not at a spot, but at any visual image projected on a screen in front of it. If this should be that of a battleship on the high seas, the bird can be trained to peck at its centre, wherever it is on the screen. Then the screen itself can be made sensitive to the place the bird pecks at, by appropriate mechanical systems, so that off-centre pecks will cause the image to move to the centre again. Now link this to the guidance systems, by conventional ailerons, or deflection of jets, for example, and the result is a perfect system for guiding a missile to its target, with the pigeon in the nose, directing – albeit unknowingly – affairs completely. It is clear that this system was, in principle, feasible. It had the great advantage that the deadly homing device, that is, the bird, could be trained to peck at an appropriate shape, whatever the differences in its appearance, such as colour, clarity, etc., caused by weather conditions. As far as is known, this pigeon-guidance system was not used in the Second World War; indeed, the reaction of the military authorities, on the one occasion when it was demonstrated to them, is reported as one of unqualified suspicion, if not incredulity! This is not altogether surprising in view of the general ignorance, partly due to the tradition of regarding them as 'dumb brutes', of the very real capacities of animals to learn and discriminate. But one might have supposed that military men would have a soft spot for the pigeon, in view of the long and honourable part played in military intelligence by carrier pigeons before the advent of the telegraph and, more especially, radio.

Our knowledge of animal behaviour can, however, be put to benign uses. A familiar example is that of training guide dogs for the blind. Here the principles of learning are used to assist in meeting some of the problems encountered in training these animals for their exacting responsibility. Thus at one guide-dog school it was found that many dogs, most of which are of the familiar Alsatian breed, were failing to reach the standard required of them as guide dogs. They were frightened of people, or unstable and aggressive, or too bothered by the correction required in their training. It was discovered that the dogs which failed the tests were those which had had a longer time in the kennels with only a small number of human contacts, before being placed in a home with children in order to get used to humans. Now it seems likely that the age at which this experience starts is crucial: delay it too long and the dog will be less likely to develop a mature 'personality', however promising it has proved to be as a young pup.

Another way animals help man is in performing for his amusement, and we are all familiar with the circus tricks in which dogs, seals, elephants, and the like go through routines which seem unbelievable, if one takes them too seriously. Remember that the bear riding the bike is not interested in bicycling as a pastime at all, but only in the reward it will get for performing the trick. Thus the circus trainer is putting into practice the principles of learning which have been our concern here. One application of these principles which has been more directly made is that of using stunts which can be built up in the form of a series of habits for the purpose of advertising displays. Thus, a group of chicks can be trained to climb a little ladder to a platform to obtain a morsel of food, then jump off – or be pushed off by the next chick in the line – and land on a plate which, when depressed in this way, makes an electric contact and lights up the trade name. The chick then runs back to the foot of the ladder and climbs again, thus repeating the process. With several chicks in this endless loop an effective display is presented; when the performers tire they can be replaced with a fresh batch which has been trained in this way but kept in reserve meanwhile. Other types of display can be achieved: examples are the pig which will choose the advertiser's product from among others on display, and the chicken which apparently switches on a miniature juke-box and 'dances' to the tune it produces before moving off to feed.

These examples of the application of our knowledge of animal behaviour are not impressive – they are not intended to be. As stressed before, this study for its own sake generates its own rewards by its intrinsic interest and the hunt for leads in discovering what makes humans 'tick'. But the future may bring changes: already there are startling developments. Some studies have shown that there are certain factory tasks which animals can do as well as, or even better than, humans. This is especially true of inspection tasks where constant alertness, keen eyesight, and dexterity are required. Humans find such tasks very boring, and the tendency has been to try to devise machines which can take over the work. Such machines are usually difficult and always costly to develop – the reason being that there are usually so many different reasons why a manufactured article may fail to reach the required specification. The size may be wrong, the colour may be wrong, it may be damaged, and so on. Different mechanized operations would be required for each type of fault, whereas a human – or an animal – inspector can take account of all of these in one operation. Moreover, there still exist many tasks for which a specialized discrimination is required, which cannot be matched by any known machine, because the possible combination of conditions of the sort mentioned and the alternatives required are too numerous or too varied to be catered for mechanically.

So far, in essaying this task, pigeons have been used. They have excellent eyesight, and the techniques of training them, mentioned several times already, are available for setting up experimental trials of their efficiency as inspectors of finished products such as transistors or pharmaceutical pills. They have been taught to examine each object as a conveyor belt takes it past a window in their cage, near which is mounted a perfect specimen of the product for comparison. Essentially, what the bird then has to do is to compare the two samples; if the test object differs in any way, it pecks one key, but if it detects no difference, then it must peck a different key to get its reward. There does not need to be a reward every time – as we saw in Chapter 3, intermittent rewards will sustain work in this type of Skinner box for long periods of time – but the keys are geared to appropriate mechanisms which reject or accept each object the bird has pronounced upon. A check can be made by sending through a known 'dud' from time to time to see if the bird will deal with it appropriately.

Now the advantage of this system is obvious. Bird labour is cheap and can be trained to whatever degree of perfection is desired. Thus the manufactured products could pass several inspectors, each more exacting than the last. As with a human inspectorate, any sort of defect can be handled by the animal – unlike a machine inspection which might be unable to handle a novel defect which had not been foreseen and which the machine had not been built to reject.

If this strikes you as fantastic, consider the possibility that birds could be trained to do postal sorting, that is, to read characters, perhaps in a code, on envelopes. 'But,' the reader may object, 'earlier you were stressing the difficulty of demonstrating a true number sense in animals and here you are talking of birds reading!' But it could be done if each bird was trained to recognize one letter, and one letter only, however it was written. Using the methods discussed previously, the bird would be trained to peck a target for a food reward whenever any variant of the selected letter appeared on a screen in front of it and to peck a different target for any other letter. Then simultaneously flash each letter in the address in turn on twenty-six screens in twenty-six Skinner boxes. Only one bird will respond positively, and a mechanism could record which one – say, bird number eight, if the first letter is 'H'. Then perhaps bird number twenty-one will respond to the next letter ('U') and then number twelve, and the letter would thus be sent into the appropriate (and, in this case most appropriately named) pigeon-hole for dispatch to the city of Hull.

For tasks needing some manipulative skill, like turning knobs, as well as a greater degree of intelligent behaviour than birds possess, monkeys, especially apes such as chimpanzees, could be used for the control of continuous processes which have to be constantly maintained. In some industries, such as chemical engineering or oil refining, the time for the admixture of certain ingredients, or the application of a particular treatment, or the precise regulation of temperature, viscosity, etc., cannot be foreseen in advance but must depend on observation of the way the process develops each time. Given that information of this kind could be presented in such a way as to be recognizable by apes, the control of such continuous processes could probably be left to them. I even think it likely that a chimpanzee could be taught to drive a train safely. Whether people would trust themselves to a non-human driver is problematical, however! Such is our faith in the machine, on the other hand, that we would, and in aircraft frequently do, trust ourselves to automatic devices of this kind.

But it is in the realm of the wider service of man in food production that animals, and especially primates, may come into their own on a scale which would make our present use of dogs and ferrets for hunting, of cormorants for fishing, and even of monkeys for harvesting coconuts, seem pitifully small.

Whole crops may be harvested by ape labour in the future. Whereas the tendency in the affluent Western societies has been to mechanize as many of such tasks as possible, in less developed countries often neither the capital to acquire such machines nor the power to operate them exists. The exploitation in the service of man of the behavioural resources of animals has hardly begun. The future may see a greatly extended application of the lowlier talents revealed by the science of animal behaviour.

Further Reading

SCOTT, J. P., *Animal Behavior*, Chicago (Chicago University Press), 1958 (for Chapter 1)

TINBERGEN, N., *Curious Naturalists*, London (Country Life), 1958 (for Chapter 2)

MUNN, N. L., *Handbook of Psychological Research on the Rat*, London (Harrap), 1950 (for Chapter 3)

THORPE, W. H., *Learning and Instinct in Animals*, London (Methuen), 1956 (for Chapters 4 and 5)

WATERS, R. E., RETHLINGSHAFER, D. A., & CALDWELL, W. E., *Principles of Comparative Psychology*, New York (McGraw-Hill), 1960 (for Chapters 4 and 5)

EYSENCK, H. J., *Handbook of Abnormal Psychology*, London (Pitman), 1960 (for Chapter 6)

Index